The Three-Body Problem and the Equations of Dynamics

Astrophysics and Space Science Library

EDITORIAL BOARD

Chairman

W. B. BURTON, *National Radio Astronomy Observatory, Charlottesville, Virginia, U.S.A. (bburton@nrao.edu); University of Leiden, The Netherlands (burton@strw.leidenuniv.nl)*

F. BERTOLA, *University of Padua, Italy*
C. J. CESARSKY, *Commission for Atomic Energy, Saclay, France*
P. EHRENFREUND, *Leiden University, The Netherlands*
O. ENGVOLD, *University of Oslo, Norway*
A. HECK, *Strasbourg Astronomical Observatory, France*
E. P. J. VAN DEN HEUVEL, *University of Amsterdam, The Netherlands*
V. M. KASPI, *McGill University, Montreal, Canada*
J. M. E. KUIJPERS, *University of Nijmegen, The Netherlands*
H. VAN DER LAAN, *University of Utrecht, The Netherlands*
P. G. MURDIN, *Institute of Astronomy, Cambridge, UK*
B. V. SOMOV, *Astronomical Institute, Moscow State University, Russia*
R. A. SUNYAEV, *Space Research Institute, Moscow, Russia*

More information about this series at http://www.springer.com/series/5664

Henri Poincaré (Deceased)

The Three-Body Problem and the Equations of Dynamics

Poincaré's Foundational Work on Dynamical Systems Theory

Translated by
Bruce D. Popp

Springer

Henri Poincaré (Deceased)
Paris
France

Translated by Bruce D. Popp

ISSN 0067-0057   ISSN 2214-7985 (electronic)
Astrophysics and Space Science Library
ISBN 978-3-319-52898-4   ISBN 978-3-319-52899-1 (eBook)
DOI 10.1007/978-3-319-52899-1

Library of Congress Control Number: 2017932019

Original Edition: Sur le problème des trois corps et les équations de la dynamique © Institut Mittag-Leffler—Sweden 1890

© Springer International Publishing AG 2017
This work is subject to copyright. All rights are reserved by the Publisher, whether the whole or part of the material is concerned, specifically the rights of translation, reprinting, reuse of illustrations, recitation, broadcasting, reproduction on microfilms or in any other physical way, and transmission or information storage and retrieval, electronic adaptation, computer software, or by similar or dissimilar methodology now known or hereafter developed.
The use of general descriptive names, registered names, trademarks, service marks, etc. in this publication does not imply, even in the absence of a specific statement, that such names are exempt from the relevant protective laws and regulations and therefore free for general use.
The publisher, the authors and the editors are safe to assume that the advice and information in this book are believed to be true and accurate at the date of publication. Neither the publisher nor the authors or the editors give a warranty, express or implied, with respect to the material contained herein or for any errors or omissions that may have been made. The publisher remains neutral with regard to jurisdictional claims in published maps and institutional affiliations.

Printed on acid-free paper

This Springer imprint is published by Springer Nature
The registered company is Springer International Publishing AG
The registered company address is: Gewerbestrasse 11, 6330 Cham, Switzerland

Translator's Preface

I first heard of this monograph—*Sur le problème des trois corps et les équations de la dynamique* by Henri Poincaré—around 1978 while I was an undergraduate at Cornell University or around 1982 while I was a graduate student in the astronomy department at Harvard University. I no longer remember the particular time or context, although there are a few conceivable possibilities.

What is clear, many years later, is that I was in an environment that recognized, respected, and understood (on some level) the importance of this monograph. And it did so despite two obstacles. The first was of course its age; it was published in 1890. The other obstacle, from the perspective of a US academic environment, was the language; it was written in formal French with a specialized vocabulary demanded by the subject matter. In 1982, I spoke French that was fully adequate for many purposes; yet it then seemed to me unlikely that I would be able to read and understand Poincaré's work, so I made no effort to try.

Together, this means that the monograph was a classic and inaccessible to a large readership even though its existence was well known.

On the way from 1985 (when I was awarded a Ph.D.) to 2014, my life and career experienced some strange twists and turns and sharp bumps and jolts. By then I had become an established, independent translator from French into English working mostly with complex technical subjects.

My command of French had improved and I had succeeded in adding to that base the complex skills of mental magic needed to translate. I had also developed the skills for researching and learning the particular terminology of specialized subject matter. In brief, I had acquired the language skills to make what had been inaccessible accessible.

One of the distressing realities of freelance work is the unpredictable switch between frenetic feeding frenzy and frustrating famine. In the spring of 2014, during one such famine, I started to look for stimulating intellectual activity to fill the time until the next feeding frenzy hit. I immediately focused my search on potential projects that could make a connection back to what I had once been: an astronomer and mathematical physicist.

In fairly short order I had a few ideas for projects involving dynamics or stability of rotating astrophysical fluids. I talked to some people. I tried to assess the effort and resources that might be needed. While this route seemed plausible, it did not grab hold of my interest and hang on.

At the same time, my interest in Henri Poincaré's work resurfaced; I was very interested in carefully understanding what Poincaré had written. What better way to do that than to translate his monograph? I quickly found that it was easy to find scanned images of his works online. (As an interested reader, it is worth your while to visit the web site hosted by the Université de Lorraine for the Henri Poincaré Papers and the bibliography in particular.) In addition to this monograph, I also looked at *Les méthodes nouvelles de la mécanique céleste* and his three books popularizing science (*La Science et l'hypothèse*, *La Valeur de la science* and *Science et méthode*). These last four books all had existing translations of unknown quality. (Not all translators are equal, far from it.) It was also clear that setting the equations in *Sur le problème* and *Les méthodes nouvelles* would require a significant effort. On the other hand, I recognized that I would likely find that effort satisfying.

Even more than that, I found Poincaré to be a compelling author.

I prepared a sample translation of a chapter from *La Science et l'hypothèse* and after discussions with Maria Ascher and Michael Fischer then at Harvard University Press, I decided to dive in and start translating *Sur le problème* motivated by my interest in the author and subject.

As paying translation work flowed in, I translated patents and documents for clinical trials and as that work ebbed, I went back to translating mathematical physics. In that way, I got two things that really interested me: stimulating intellectual activity, and close, detailed study of a monograph and author that had long interested me.

What you have before your eyes is the product of a labor of love.

The Monograph

This monograph is the foundational work in the theory of dynamical systems.

Before this work, it was generally thought that the motion of the planets in the solar system proceeded in orderly fashion, like a clockwork orrery, from a known past to a predictable future since their motion was governed by deterministic laws: Newton's laws of gravitation and motion. Work on the computational machinery for predicting planetary positions into the future had progressed through the nineteenth century with notable contributions from Joseph-Louis Lagrange, Pierre-Simon Laplace and Urbain Le Verrier. The discovery of Neptune by Johann Galle within $1°$ of the position predicted by Le Verrier was a major triumph.

The computational machinery for predicting future positions of planets under these deterministic laws involves the use of time series expansions. The practical value of these time series, at least over some time domains, was well established.

The mathematical convergence of the series predicting planetary positions for all time was unproven. Phrased another way, it was unknown whether the orbits of the planets in the solar system were stable for all time.

Providing this proof was the subject of a prize competition celebrating the 60th birthday of King Oscar II of Sweden in 1889. This monograph is the published result of Poincaré's entry in that prize competition. Readers wishing more information about the prize competition and the history surrounding this monograph would do well to consult the book by June Barrow-Green (Barrow-Green 1997).

To approach the prize problem, Poincaré did not start with the solutions for the planetary positions. Instead he approached from the other direction by studying the equations of dynamics in general, Hamiltonian form. In that way, he developed the tools and theorems for studying the equations of dynamics in order to understand the behavior of their solutions. The immediate result was many tools and theorems that are now central to dynamical systems theory: phase space, trajectories, recurrence, Poincaré maps and more. (See, at the end of this Preface, the *Notable Concepts* section).

Hilborn writes (Hilborn 2000) on page 62, "Even a cursory reading of the history of chaos, the definitive version of which is not yet to be written, shows that Poincaré knew about, at least in rough way, most of the crucial ideas of nonlinear dynamics and chaos." And on the same page, he continues, "It is safe to say, however, that if Poincaré had a Macintosh or IBM personal computer, then the field of nonlinear dynamics would be much further along in its development than it is today." This is a very interesting anachronism.

With this theory of dynamical systems, Poincaré proved that the convergence of the solutions of the equations of dynamics cannot be established in general for all time. The series expansions may converge and be useful in some cases and for some time domains, and diverge over longer time. We do not know, and we cannot know, whether the solar system is stable.

Anyone interested in the history of chaos or in understanding how well and completely Poincaré understood the crucial ideas of dynamical systems theory will find it rewarding to explore this monograph. Beyond my personal interest, I prepared this translation so you could explore this monograph for yourself. In (Lorenz 1993), Lorenz asks on p. 118, "Did [Poincaré] recognize the phenomenon of full chaos, where most solutions—not just special ones—are sensitively dependent and lack periodicity?" Reading Chap. 4, *Encounters with Chaos*, of (Lorenz 1993) I wonder how much (well, actually, how little) Lorenz knew of the content of Poincaré's work in this monograph and in *Les méthodes nouvelles de la mécanique céleste*. Only *Les méthodes nouvelles* and the last of the three popularizations are cited in his Bibliography. With this translation, you can read through this work by Poincaré for yourself to form your own informed answer.

Errors and Typos

In the Preface, H. Poincaré writes, "By drawing my attention to a delicate point, he enabled me to discover and rectify a significant error." Not only was the error significant, the timing was bad. As the sentence suggests, the error was found and corrected without knowledge of it becoming public and the correction and reprinting resulted in a substantially different monograph becoming available to the public (Barrow-Green 1997).

Despite this challenging publication history, the published version of the monograph that reached the public has a limited amount of lint, distracting errors of a typographic nature not affecting the fabric of the work. I have found 26. For example, on pages 32–33 the equation numbers advance from 3 to 5 and equation 4 does not appear anywhere else in this section. There is nothing to be done about an error like this during translation and so the error is repeated. On the other hand, on page 34 in equation 6, the first subscript x_n is incorrect and is easily corrected to x_1. In circumstances like this I have corrected the errors unobtrusively.

I firmly hope that I have not contributed to the lint (or worse) and I have been diligent in my efforts to avoid (or failing that to find and correct) errors large and small of my own creation.

The Translation

In preparing this translation, I have tried to keep before me several objectives.

The first is accessibility. At one level, this objective is true of any translation. The purpose of translation is to take a document which was written (and therefore accessible) in one language and fit for a particular purpose and render it in another language (and therefore accessible in that language too) where it is fit for the same purpose or some analog of that purpose. In this instance, I understand that purpose to be a scholarly presentation of Poincaré's ideas and approach to studying and understanding dynamical systems and particularly the general three-body problem. Implicit here are the ideas of time and audience: 125 years later the expected audience for this translation is English-speaking people knowledgeable in dynamical systems wishing to understand how a foundational classic of the field established and set its direction.

Looking deeper, there is also the issue of voice. In the translation, in contrast to this preface, I have tried to avoid speaking for myself, meaning retelling in my words what Poincaré wrote, and to follow closely what and how Poincaré wrote, meaning to let his voice come through, while respecting standards of grammar, syntax, and phrasing expected in contemporary professional US English.

Essential to both of these is the matter of accuracy. In preparing this translation, I have worked through and sought to understand what Poincaré was writing about so that I would be able to accurately present it in my translation. I have then checked

Translator's Preface ix

and rechecked this translation to eliminate any misunderstanding, inconsistency, or infelicity that might have gotten through anyway. I am human so I can be certain that I have not been fully successful despite my best effort.

Fundamental to my effort and motivation is my opinion that this is a classic of our literature in the field that deserves to be understood and that Poincaré merits the recognition and credit that follows from that understanding.

References and Index

Poincaré provided in line references. In some cases the reference amounted to little more than a name and a subject area. The least detailed citation was, "it is useful to cite the work of Mr. Puiseux on the roots of algebraic equations." He did not provide an index.

A list of references Poincaré cited (including the above example) and an index have been prepared to accompany this translation. They can be found after the translation.

Notation

As a consequence of the objective of helping Poincaré's voice to come through clearly, in the sentences and words and in the equations and symbols, there are several considerations concerning his choice of notation and symbols which should be indicated in order to help the reader appreciate the content.

Vectors

Here Poincaré does not use vector notation. He consequently writes things like "x_1, x_2 and x_3" and "the x" where we would expect to see $\bar{x}$. Reading his work with this in mind, you will easily recognize places where equations could be written more compactly with vectors and connected notational machinery such as dot products. After some experience reading his equations and following his reasoning, I became comfortable with the use of subscripts and summations and could see that it presented some advantages.

Superscripts as Indices

Related, and perhaps a consequence of using subscripts to indicate components of a vector, is the use of superscripts to indicate the order of the term in a series expansion. Therefore, x_1^2 is the second term in a series expansion of the first component of the coordinate x. There is certainly a possibility for confusion with the square of the first component. In a footnote on page 96 Poincaré warns that these "are indices and not exponents." Now I have warned you too.

Full and Partial Derivatives

In the text Poincaré refers to partial derivatives and partial differential equations 11 times. In his equations he makes no distinction between full and partial derivatives and uses d consistently for both.

As an example consider equation 1 on page 5 which Poincaré writes:

$$\frac{dx_i}{dt} = \frac{dF}{dy_i}, \quad \frac{dy_i}{dt} = -\frac{dF}{dx_i}. \tag{1}$$

In this case only the derivatives with respect to time are full derivatives; in contrast the derivatives on the right-hand sides are partial derivatives. Using ∂ to indicate partial derivatives, these equations can be rewritten

$$\frac{dx_i}{dt} = \frac{\partial F}{\partial y_i}, \quad \frac{dy_i}{dt} = -\frac{\partial F}{\partial x_i}.$$

As a plausible rule of thumb, you can assume the derivatives with respect to time are full derivatives and derivatives with respect to other variables are partial derivatives. Poincaré does make frequent use of transformations and substitutions of variables and this can cause some uncertainty as to how a new variable is related to a variable from the equations in canonical form. In case of doubt about whether a specific derivative should be a full or partial derivative, the interested reader will need to work through the substitutions and changes made to the equations in canonical form as given above.

Hamiltonian

Poincaré does not mention the name Hamilton. Poincaré refers to the canonical form of the equations of dynamics or simply the canonical equations. What Poincaré calls the canonical equations are given by equation 1 from page 5 (among other places), repeated above.

With a simple comparison, the reader can verify that these are in fact Hamilton's equations. This involves identifying F with the Hamiltonian H, the x_i (linear variables in Poincaré's terminology) with the generalized coordinates q_i and the y_i (angular variables in Poincaré's terminology) with the generalized momenta p_i. This allows writing the canonical equations in a more familiar form:

$$\frac{dq_i}{dt} = \frac{\partial H}{\partial p_i}, \quad \frac{dp_i}{dt} = -\frac{\partial H}{\partial q_i}.$$

The Hamiltonian is the total energy and Poincaré notes on page 4 that it is conserved.

Translator's Preface　　　　　　　　　　　　　　　　　　　　　　　　　　　　xi

An important source of generality in Poincaré's results is his reliance on a generic Hamiltonian. In fact the most important constraint that Poincaré places on the Hamiltonian (see for example page 90) is that it can be expanded in a power series of a small parameter μ. In the case of the three-body problem, the small parameter is identified with the reduced mass of the third body. If the third body is massless, then $\mu = 0$. He is sparing in his reference to a specific form of the Hamiltonian; his discussion of the work of G.W. Hill starting on page 63 is a rare example of explicitly stating the form of the Hamiltonian.

Section 4 of Chapter 2, Theorem III

In his proof of Theorem III on pages 74–76, Poincaré references four scenarios and provides two figures, Figs. 1 and 2. A reader who follows the proof will recognize that Scenario 2 corresponds to Fig. 1 and Scenario 3 corresponds to Fig. 2. For the reader's convenience, the following two sketches show Scenario 1 and the two variants of Scenario 4.

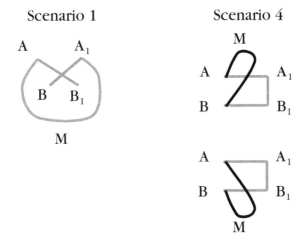

Notable Concepts

This section is provided as an aid to readers who, after consulting the table of contents, want a further indication where certain concepts are applied or discussed by Poincaré. It is therefore my effort to help readers find what they might be looking for or to make them aware of interesting things they might not have expected to find.

While this may seem similar to an index, it should be noted that unlike an index there is no effort to be comprehensive either in selecting concepts or in identifying all the places were the selected concepts occur. The discussion here, when there is any, is deliberately kept brief since the reader has access directly to Poincaré's own words for expansion and discussion.

This selection inherently reflects my background and therefore my ability to recognize certain concepts; other readers of Poincaré's work will make discoveries matching their interests.

Asymptotic solutions

p. 124: Definition

Asymptotic stability; Stability in the meaning of Poisson

p. 58: Definition

Asymptotic surfaces

p. 125, 161: Definition

Characteristic exponents

Section 2 of Chap. 3
p. 84: Definition
p. 86–87: For dynamical systems, they are equal pairwise and of opposite sign.
p. 108: Proof they can be expanded in $\sqrt{\mu}$

Collisionless

p. 11–12: Collisions result in singular points in Newton's law of gravitation preventing convergence of series expansions. The problems considered must therefore be collisionless.

Commensurable mean motions

p. 90–91: The existence of periodic solutions of the three-body problem requires that the mean motions of the three bodies be commensurable. See also *Small Divisors*.

Contactless surface

p. 56: Definition
p. 73: Lemma II

Delaunay variables (as used by Tisserand)

p. 151: Introduced. Relationships to semi-major axis, eccentricity and mean motion are given. With these variables, the equations for the orbital elements can be written as canonical equations of dynamics.

Density of trajectories

p. 58: Theorem I
p. 60: Corollary
p. 212: In a region of libration

Translator's Preface

Doubly asymptotic (now known as homoclinic) trajectories

p. 200: Definition

Duffing's equation

p. 103 (equation 18) and 157 (unnumbered): As written, this form of the equation is undamped and weakly forced. Poincaré provides a change of variables that allows the equation to be rewritten in the canonical form of the equations of dynamics.

Existence of periodic solutions for small μ

p. 91–94: Proof

G.W. Hill

p. 63–64: Discussed in terms of integral invariants, bounded solutions.
p. 65–68: Results extended by Poincaré and Bohlin.
p. 68–69: Hill's bounded solutions for the restricted three-body problem do not extend to the general three-body problem.

Hamiltonian, explicit forms

p. 41: Provides definition in terms of kinetic and potential energy
p. 51: For inverse square law
p. 63: G. W. Hill
p. 103 and 157: For Duffing's equation
p. 151: In terms of *Delaunay variables*
p. 196: Weakly coupled pendulum and spring, used as a counterexample. (For further discussion see (Holmes 1990) section 5 where this is equation 5.1 with $\varphi(y)$ (Poincaré) identified with $\sin q_1$ (Holmes). Poincaré assumes $\varphi(y) = \sin y$ on page 198.)

Homoclinic trajectories

See *Doubly asymptotic trajectories*.

Hyperboloid

p. 163: An asymptotic surface with two sheets

Infinitesimal contact transformation; Perturbation of Hamilton's equations

p. 37–38: A first-order Taylor series expansion of Hamilton's equations near a known solution is used to study a second infinitesimally close solution. This leads to relationships between the two solutions.

Integral Invariants

Section 2 of Chap. 2
p. 43: Definition.
p. 44–45: Proof of Poincaré's theorem.

Invariant Curve

p. 69–70: Definition

Libration

p. 206: Orbital libration near Lagrange points.
p. 209–212: In a region of libration, there are infinitely many closed trajectories corresponding to periodic solutions.

Linear and angular variables

p. 6: Definition. These take the place of coordinates and conjugate momentums.

Lunar problem

p. 150: Stated

Lindstedt

Last paragraph of Poincaré's Preface, p. 69 and Sect. 2 of Chap. 6

Liouville's Theorem

p. 45–46: Discussion; Volume in phase space of an incompressible fluid is an integral invariant.
p. 47–49: Proof

Newton's Law of Gravitation

p. 11

Orbital Elements

See *Delaunay variables*.
p. 90: Defines mean motion (n_i), longitude of pericenter (ϖ_i)

Periodic Forcing

p. 86: The assumptions allow either a time independent Hamiltonian or periodic forcing with period 2π.

Periodic Solutions of the First Kind

p. 89: Proof of the existence of solutions of the equations of dynamics for $\mu = 0$ which continue to exist as periodic solutions for small values of μ.

Periodic Solutions of the Second Kind

Section 1 of Chap. 6: These solutions do not exist for $\mu = 0$, but do exist for $\mu > 0$.
p. 212: Periodic solutions of the second kind disappear as μ decrease continuously to 0.

Poincaré map

p. 56: Discusses a point P_0 where a trajectory intersects a portion of surface S and the recurrences $(P_1, P_2, \ldots P_n)$ of that point.

Poincaré's theorem

p. 45–46: Proof

Poisson Bracket

p. 40: Defined (but not explicitly named).
p. 51: Used in proof that a quantity is an integral invariant.

Poisson Stability

p. 58: Definition

Poisson's Theorem

p. 40: The Poisson bracket of two, constant integrals of the canonical equations is a constant.

Recurrence

p. 56: Definition

Recurrence and Integral Invariants

p. 57: Extends integral invariants to areas related by recurrence.

Recurrence Theorem

p. 58: Theorem I

Small divisors

p. 239: Avoided by commensurability relation between mean motions. See also *Commensurable Mean Motions*

State Space (now commonly called phase space)

p. 5: No explicit definition or discussion, but used throughout. See p. 6, "The state of the system could then be represented by a point." For higher dimensions, visualization is a problem (p. 6) and one is dependent on "analytical language" (p. 69).

Temporary (linear) Stability

p. 88: Definition

Testing convergence of series (Cauchy, Weierstass, Kowalevski, Poincaré)

Section 2 of Chap. 1: Testing convergence by term-by-term comparison of series in question to a converging, reference series.
Section 3 of Chap. 1: Extension by Kowalevski (variously spelled, including Kovalevskaya) and Poincaré.

Tisserand relation

p. 152, unnumbered equation: The quantity $F_0 = 1/2a + G$ is the zeroth order term in the expansion in μ of the Hamiltonian for the planar (zero inclination) three-body problem. The Tisserand relation states that F_0 is unchanged after a close encounter of a comet with Jupiter. The full Hamiltonian is an integral invariant and for very small μ (for example a comet or spacecraft coming close to Jupiter) F_0 is very nearly an integral invariant.

Total Energy is an integral invariant

p. 43–44: Proof

Total energy along an arc in phase space is an integral invariant

p. 51–52: Proof

Trajectory

p. 5: Definition

Trajectory surface

p. 56–57: Definition

Acknowledgements

Several people provided particular encouragement or useful discussions. I would like to acknowledge the support and advice I received from friends and family—Lora Lunt, Cathy Thygesen, Greg Farino, Fred Marton, Miriam Popp, Andrew Popp, Cristina Aldrich and Mireille Popp—my translation colleagues—Maria Ascher, Marian Comenetz, Ken Kronenberg and Judy Lyons—and academic mentors—Peter Gierasch, Owen Gingerich, and Bob Rosner. Thank you.

Le Laboratoire d'Histoire des Sciences et de Philosophie - Archives Henri-Poincaré. CNRS/Université de Lorraine. http://henripoincarepapers.univ-lorraine.fr allowed the use of the images of the original figures from their scanned copy.

Norwood, MA, USA   Bruce D. Popp
December 2016

Bibliography

Barrow-Green, J. (1997). *Poincaré and the Three Body Problem.* Providence, Rhode Island: American Mathematical Society.

Hilborn, R. C. (2000). *Chaos and Nonlinear Dynamics, Second Edition.* New York: Oxford University Press.

Holmes, P. (1990). Poincaré, Celestial Mechanics, Dynamical-Systems Theory and "Chaos". *Physics Reports (Review Section of Physics Letters), 193* (No. 3), 137–163.

Lorenz, E. N. (1993). *The Essence of Chaos.* Seattle, WA: University of Washington Press.

Author's Preface

This work studying the three-body problem is a reworking of the monograph that I had presented to the Competition for the prize established by His Majesty the King of Sweden. This reworking had become necessary for several reasons. Short on time, I had needed to present several results without proof; with the help of the indications which I had provided, the reader would have been able to reconstruct the proofs only with great difficulty. I had first considered publishing the initial text and accompanying it with explanatory notes; the result was a multiplication of these notes and reading the monograph would have been demanding and unpleasant.

I therefore preferred to merge these notes into the body of the work and this merging had the advantage of avoiding some restatements and better positioning the ideas in logical order.

I must heartily acknowledge the contribution of Mr. Phragmén who not only reviewed the proofs with great care, but who—having read the monograph with attention and having penetrated the meaning with utmost finesse—showed me places where he thought additional explanations were necessary to bring out the full insight of my thoughts. The elegant form that I gave to the calculation of S_i^m and T_i^m at the end of §12 is due to him. By drawing my attention to a delicate point, he enabled me to discover and rectify a significant error.

In some of the additions that I made to the initial monograph, I limited myself to reviewing some previously known results. Since these results are scattered in many works and I had made frequent use of them, I thought the reader would be well served by sparing them tedious searches; furthermore, I was often led to apply these theorems in a different form than first given to them by their author and it was indispensable to present them in this new form. In Chap. 1 (Part I), these existing theorems—some of which are even classics—are developed side-by-side with some new propositions.

I am still a long ways from having fully resolved the problem that I have taken on. I have limited myself to demonstrating the existence of some remarkable specific solutions which I call periodic solutions, asymptotic solutions and doubly asymptotic solutions. I have more closely studied a particular case of the three-body problem in which one of the masses is zero and the motion of the two others is circular. I have recognized that in this case the three bodies will return arbitrarily close to their initial position infinitely many times, unless the initial conditions of the motion are exceptional.

As can be seen, these results show us very little about the problem's general case, but they could have some worth, because they are rigorously established, whereas the three-body problem had until now seemed accessible only by methods of successive approximation where the absolute rigor, which is required in other areas of mathematics, is given away cheaply.

But I would especially like to draw the reader's attention to the negative results which are developed at the end of the monograph. For example, I established that apart from known integrals the three-body problem does not comprise any analytic and one-to-one integral. Many other circumstances lead us to expect that the full solution, if it can ever be discovered, will demand analytical tools which are absolutely different from those which we have and which are infinitely more complicated. The more thought given to the propositions that I demonstrate later on, the better it will be understood that this problem was incredible difficulties that has certainly suggested by the lack of success in prior efforts, but I think I have brought out the nature and the immensity even better.

I also show that most of the series used in celestial mechanics and in particular those of Mr. Lindstedt, which are the simplest, are not convergent. I am sorry in that way to have thrown some discredit on the work of Mr. Lindstedt or on the more detailed work of Mr. Gyldén; nothing could be farther from my thoughts. The methods that they are proposing retain all their practical value. In fact the value that can be drawn from a numeric calculation using divergent series is known and the famous Stirling series is a striking example. Because of an analogous circumstance, the tried-and-true developments in celestial mechanics have already rendered such great service and are called on to render even greater service.

One of the series, which I will make use of later and whose divergence I will furthermore prove, has a significant analogy with the development proposed to the Stockholm Academy May 9, 1888 by Mr. Bohlin. Since his monograph was printed a few months later, I was unaware of it at the time the competition closed, meaning June 1, 1888. That means that I did not cite Mr. Bohlin's name and I am making an effort here to give him the justice he deserves. (See Supplement to the minutes of the Stockholm Academy, Volume 14 and Astronomische Nachrichten, Number 2883.)[1]

Paris, France

Henri Poincaré

---

[1]Translator: It appears that the reference should be to number 2882, *Zur Frage der Convergenz der Reihenentwickelungen in der Störungstheorie*.

Contents

Part I Review

1 General Properties of the Differential Equations 3
   1 Notations and Definitions 3
   2 Calculation of Limits 7
   3 Application of the Calculation of Limits to Partial Differential
      Equations 18
   4 Integration of Linear Equations with Periodic Coefficients 31

2 Theory of Integral Invariants 37
   1 Various Properties of the Equations of Dynamics 37
   2 Definitions of Integral Invariants 43
   3 Transformation of Integral Invariants 53
   4 Using Integral Invariants 57

3 Theory of Periodic Solutions 77
   1 Existence of Periodic Solutions 77
   2 Characteristic Exponents 84
   3 Periodic Solutions of the Equations of Dynamics 89
   4 Calculation of the Characteristic Exponents 106
   5 Asymptotic Solutions 119
   6 Asymptotic Solutions of the Equations of Dynamics 127

Part II Equations of Dynamics and the N-Body Problem

4 Study of the Case with Only Two Degrees of Freedom 147
   1 Various Geometric Representations 147

5 Study of the Asymptotic Surfaces 161
   1 Description of the Problem 161
   2 First Approximation 163

3	Second Approximation	174
4	Third Approximation	195

6 Various Results 203
   1 Periodic Solutions of the Second Kind 203
   2 Divergence of Lindstedt's Series 222
   3 Nonexistence of One-to-One Integrals 230

7 Attempts at Generalization 237
   1 The N-Body Problem 237

Erratum 241

Bibliography 243

Author Index 245

Subject Index 247

Part I
Review

Chapter 1
General Properties of the Differential Equations

1 Notations and Definitions

Consider a system of differential equations:

$$\frac{dx_1}{dt} = X_1, \quad \frac{dx_2}{dt} = X_2, \quad \ldots \frac{dx_n}{dt} = X_n, \tag{1}$$

where t represents the independent variable that we will call time and $x_1, x_2, \ldots x_n$ the unknown functions, where $X_1, X_2, \ldots X_n$ are given functions of $x_1, x_2, \ldots x_n$. We suppose in general that the functions $X_1, X_2, \ldots X_n$ are analytic and one-to-one for all real values of $x_1, x_2, \ldots x_n$.

If one knew how to integrate Eq. (1), the result of the integration could be stated in two different forms; we could write:

$$\begin{aligned} x_1 = \varphi_1(t, C_1, C_2, \ldots C_n), \quad x_2 = \varphi_2(t, C_1, C_2, \ldots C_n), \\ \ldots x_n = \varphi_n(t, C_1, C_2, \ldots C_n), \end{aligned} \tag{2}$$

where $C_1, C_2, \ldots C_n$ designate the constants of integration.

This could also be written, by solving for these constants:

$$\begin{aligned} C_1 &= F_1(t, x_1, x_2, \ldots x_n), \\ C_2 &= F_2(t, x_1, x_2, \ldots x_n), \\ &\vdots \\ C_n &= F_n(t, x_1, x_2, \ldots x_n). \end{aligned} \tag{3}$$

To avoid any confusion, we will say that if the constants C remain arbitrary then Eq. (2) represents the *general solution* of Eq. (1) and that if numerical values are given to C than they represent a *specific solution*. We will also say that in Eq. (3),

$F_1, F_2, \ldots F_n$ are n *specific integrals* of Eq. (1). The meaning of the word *solution* and *integral* are thereby completely set.

Suppose that a specific solution of Eq. (1) is known that can be written:

$$x_1 = \varphi_1(t), \quad x_2 = \varphi_2(t), \quad \ldots \quad x_n = \varphi_n(t). \tag{4}$$

The study of specific solutions of (1) that are slightly different from the solution (1) can be proposed. To do that set:

$$x_1 = \varphi_1 + \xi_1, \quad x_2 = \varphi_2 + \xi_2, \quad \ldots \quad x_n = \varphi_n + \xi_n$$

and take $\xi_1, \xi_2, \ldots \xi_n$ for new unknown functions. If the solutions that we want to study are slightly different from the solution (4), then the ξ are very small and we can neglect their squares. Equation (1) then becomes, by neglecting higher powers of ξ:

$$\frac{d\xi_i}{dt} = \frac{dX_i}{dx_1}\xi_1 + \frac{dX_i}{dx_2}\xi_2 + \cdots + \frac{dX_i}{dx_n}\xi_n \quad (i = 1, 2, \ldots n) \tag{5}$$

In the derivatives dX_i/dx_k, the quantities $x_1, x_2, \ldots x_n$ must be replaced by $\varphi_1(t), \varphi_2(t), \ldots \varphi_n(t)$, such derivatives can be regarded as known functions of time.

Equation (5) is *first-order Taylor-series expansions* and will be called perturbation equations of Eq. (1). It is seen that these equations are linear.

Equation (1) is called *canonical* when there are an even number of variables $n = 2p$, separating into two series:

$$x_1, x_2, \ldots x_p, y_1, y_2, \ldots y_p,$$

and Eq. (1) can be written:

$$\frac{dx_i}{dt} = \frac{dF}{dy_i}, \quad \frac{dy_i}{dt} = -\frac{dF}{dx_i} \quad (i = 1, 2, \ldots p).$$

They then have the form of the equations of dynamics and following the usage of English speakers we will say the system of Eq. (6) comprises p *degrees of freedom*.

This system (6) is known to allow an integral known as conservation of total energy:

$$F = \text{const.}$$

and if $p - 1$ others are known, the canonical equations can be considered to be fully integrated.

In particular, consider the case $n = 3$; we can then regard x_1, x_2, and x_3 as the coordinates of a point P in space. The equations:

$$\frac{dx_1}{dt} = X_1, \quad \frac{dx_2}{dt} = X_2, \quad \frac{dx_3}{dt} = X_3 \tag{6}$$

then define the velocity of this point P as a function of its coordinates. Consider a specific solution of Eq. (1):

$$x_1 = \varphi_1(t), \quad x_2 = \varphi_2(t), \quad x_3 = \varphi_3(t).$$

As we vary the time t, the point P will describe some curve in space; we will call it a *trajectory*. To each specific solution of Eq. (1) there therefore corresponds one trajectory and reciprocally.

If the functions X_1, X_2, and X_3 are one-to-one, then one trajectory and only one passes through each point in space. There is only an exception if one of these three functions becomes infinite or if all three are zero. The points where these exceptions occur are called *singular points*.

Consider an arbitrary skew curve. A trajectory passes through each of the points of this curve; the family of these trajectories constitutes a surface that I will call *trajectory-surface*.

Since two trajectories cannot cross except at a singular point, no trajectory can cross a trajectory-surface that does not pass through a singular point.

In the following, we will frequently need to be concerned with the question of stability. There will be *stability*, if the three quantities x_1, x_2, and x_3 remain less than certain bounds when the time t varies from $-\infty$ to $+\infty$; or in other words, if the trajectory of the point P remains entirely in a bounded region of space.

Suppose that there is a closed trajectory-surface S; this surface splits space into two regions—one inside and the other outside—and no trajectory can pass from one of these regions into the other. This means that if the position of the point P is initially in the inside region, this point will remain there forever; its trajectory will be wholly inside of S. There will therefore be stability.

The question of stability therefore amounts to the search for closed trajectory-surfaces.

This mode of geometric representation can be varied; suppose, for example, that one sets:

$$\begin{aligned} x_1 &= \psi_1(z_1, z_2, z_3), \\ x_2 &= \psi_2(z_1, z_2, z_3), \\ x_3 &= \psi_3(z_1, z_2, z_3), \end{aligned}$$

where the ψ are functions of z that are one-to-one for all real values of z. We will no longer consider x_1, x_2, and x_3, but instead z_1, z_2, and z_3, as the coordinates of a point in space. When the position of this point is known, z_1, z_2, and z_3 will be known and consequently x_1, x_2, and x_3. Everything we said above remains correct.

It is even sufficient that the three functions ψ remain one-to-one in a certain domain, provided that one does not leave this domain.

If $n > 3$, this mode of representation can no longer be used in general, unless one can manage to visualize space in more than three dimensions. There is, however, one case where this difficulty can be avoided.

Suppose for example that $n = 4$ and that one of the integrals of Eq. (1) is known. Let:

$$F(x_1, x_2, x_3, x_4) = C \qquad (7)$$

be this integral. We will regard this constant of integration C as given from the question. We can then derive one of the four quantities x_1, x_2, x_3, and x_4 from Eq. (7) as a function of the three others, or else find three auxiliary variables z_1, z_2, and z_3 such that by setting:

$$x_1 = \psi_1(z_1, z_2, z_3), \quad x_3 = \psi_3(z_1, z_2, z_3),$$
$$x_2 = \psi_2(z_1, z_2, z_3), \quad x_4 = \psi_4(z_1, z_2, z_3),$$

Equation (7) is satisfied whatever the values of z_1, z_2, and z_3. It often happens that these auxiliary variables z can be chosen such that the four functions ψ are one-to-one, if not for all real values of z, at least for a domain that it would not be necessary to leave.

The state of the system could then be represented by a point whose spatial coordinates are z_1, z_2, and z_3.

As an example, suppose we have the canonical equations with two degrees of freedom:

$$\frac{dx_1}{dt} = \frac{dF}{dy_1}, \quad \frac{dx_2}{dt} = \frac{dF}{dy_2},$$
$$\frac{dy_1}{dt} = -\frac{dF}{dx_1}, \quad \frac{dy_2}{dt} = -\frac{dF}{dx_2}.$$

We will have four variables x_1, x_2, y_1, y_2, but these variables will be related by the energy conservation equation:

$$F = C,$$

such that if we regard the conservation constant C as known, there will then only be three independent variables and geometric representation will be possible.

Among the variables $x_1, x_2, \ldots x_n$, we can recognize *linear* variables and *angular* variables. It can happen that the functions $X_1, X_2, \ldots X_n$ are all periodic with respect to one of these variables x_i and do not change when this variable increases by 2π. The variable x_i and those which display the same property will then be *angular*; the others will be *linear*.

I will state that the state of the system has not changed if all the angular variables have increased by a multiple of 2π and if all the linear values have returned to their initial values.

1 Notations and Definitions

We will then adopt a representation method such that the representative point P returns to the same point in space when one or more angular variables will have increased by 2π. We will see examples of this later.

Among the specific solutions of Eq. (1), we will distinguish the *periodic solutions*. Let

$$x_1 = \varphi_1(t), \quad x_2 = \varphi_2(t), \quad \ldots \quad x_n = \varphi_n(t)$$

be a specific solution of Eq. (1). Suppose that there exists a quantity h such that:

$$\varphi_i(t+h) = \varphi_i(t)$$

when x_i is a linear variable and

$$\varphi_i(t+h) = \varphi_i(t) + 2k\pi, \quad \text{(where is } k \text{ an integer)}$$

when x_i is an angular variable. We will then state that the solution under consideration is *periodic* and that h is the period.

If the geometric mode of representation is adopted such that the representative point remains the same when one of the angular variables increases by 2π, any periodic solution will be represented by a closed trajectory.

2 Calculation of Limits

One of Cauchy's most beautiful discoveries (*Comptes Rendus*, volume 14, page 1020), however little it might have been remarked during his time, is the one that he called the calculation of limits and for which we keep this name however poorly justified it might be.

Consider a system of differential equations:

$$\begin{aligned}\frac{dy}{dx} &= f_1(x,y,z), \\ \frac{dz}{dx} &= f_2(x,y,z).\end{aligned} \qquad (1)$$

If f_1 and f_2 are expandable according to increasing powers of x, y, and z, these equations will allow a solution of the following form:

$$y = \varphi_1(x), \quad z = \varphi_2(x),$$

where φ_1 and φ_2 are series expanded in increasing powers of x and that become zero with x.

To prove it, Cauchy replaced the two functions f_1 and f_2 by an expression of the form:

$$f'(x,y,z) = \frac{M}{(1-\alpha x)(1-\beta y)(1-\gamma z)},$$

by choosing $M, \alpha, \beta,$ and γ such that each term of f' has a larger coefficient (in absolute value) than the corresponding term of f_1 and f_2. By thus replacing f_1 and f_2 by f', the coefficients of φ_1 and φ_2 are increased and as these two series are convergent after this replacement, they would have also been convergent before this replacement.

This is the fundamental principle of the calculation of limits. Cauchy went on to make many other applications and other mathematicians have notably improved it since then.

The most important of these improvements is due to Weierstrass who replaced the function $f'(x,y,z)$ used by Cauchy by another simpler one which plays the same role.

Write Eq. (1) in the form:

$$\frac{dy}{dt} = f_1(x,y,z),$$
$$\frac{dz}{dt} = f_2(x,y,z), \qquad (1')$$
$$\frac{dx}{dt} = f(x,y,z) = 1.$$

Then replace the functions $f, f_1,$ and f_2 by the function Weierstrass used

$$f'(x,y,z) = \frac{M}{1-\alpha(x+y+z)};$$

and they become:

$$\frac{dx}{dt} = \frac{dy}{dt} = \frac{dz}{dt} = \frac{M}{1-\alpha(x+y+z)}. \qquad (2')$$

Equation (1') is formally satisfied by series:

$$x = \varphi(t) = t, \quad y = \varphi_1(t), \quad z = \varphi_2(t)$$

expanded in increasing powers of t and approaching zero with t.

Similarly Eq. (2') will be satisfied by the series

$$x = \varphi'(t) = t, \quad y = \varphi'_1(t), \quad z = \varphi'_2(t)$$

2 Calculation of Limits

expanded in a series with increasing powers of t and approaching zero with t. (It can be easily seen that $\varphi'(t) = \varphi'_1(t) = \varphi'_2(t)$).

If M and α are suitably chosen, the coefficients of the series φ' are larger than those of the series φ; now the series φ' converge; therefore the series φ also converge.
Which was to be proved.

I am not emphasizing these proofs which have become completely standard and which have been developed in all moderately complete treatments of analysis, for example in the *Cours d'Analyse* by Jordan (vol. 3, page 87).

But it is possible to go farther.

Theorem I *Imagine that the functions f_1 and f_2 depend, not only on x, y, and z, but on some arbitrary parameter μ and that they are expandable in increasing powers of x, y, z, and μ. Then, write Eq. (1) in the form*:

$$\frac{dx}{dt} = f(x,y,z,\mu) = 1, \quad \frac{dy}{dt} = f_1(x,y,z,\mu), \quad \frac{dz}{dt} = f_2(x,y,z,\mu). \tag{1''}$$

Three series can be found:

$$x = \varphi(t,\mu,x_0,y_0,z_0) = t + x_0, \quad y = \varphi_1(t,\mu,x_0,y_0,z_0), \quad z = \varphi_2(t,\mu,x_0,y_0,z_0)$$

which formally satisfy Eq. (1''), which is expanded in increasing powers of t, μ, and three integration constants x_0, y_0, and z_0 and which finally reduce, respectively, to x_0, y_0, and z_0 for $t = 0$.

I state that these series converge provided that t, μ, x_0, y_0, and z_0 are sufficiently small.

In fact, replace f, f_1, and f_2 with the function:

$$f'(x,y,z,\mu) = \frac{M}{(1-\beta\mu)[1-\alpha(x+y+z)]}.$$

This function f' is expandable in powers of x, y, z, and μ. M, α, and β can be taken sufficiently large such that each term of f' is larger than the corresponding term of f, f_1, and f_2.

We will in that way obtain the equations

$$\frac{dx}{dt} = \frac{dy}{dt} = \frac{dz}{dt} = \frac{M}{(1-\beta\mu)[1-\alpha(x+y+z)]}. \tag{2''}$$

Three series can be found

$$x = \varphi'(t,\mu,x_0,y_0,z_0), \quad y = \varphi'_1(t,\mu,x_0,y_0,z_0), \quad z = \varphi'_2(t,\mu,x_0,y_0,z_0)$$

expanded in powers of t, μ, x_0, y_0, and z_0 satisfying Eq. (2'') and reducing, respectively, to x_0, y_0, and z_0 for $t = 0$.

By reasoning as Cauchy did, it can be demonstrated that each term of the series φ' is larger than the corresponding term of the series φ. However, the series φ' converges if t, μ, x_0, y_0, and z_0 are sufficiently small. Therefore the series φ also converges. Which was to be proved.

Various conclusions can be drawn from this.

Theorem II *We just saw that x, y, and z are* expandable *in powers of the* four variables μ, x_0, y_0, *and* z_0 *provided that these five variables, including t, are sufficiently small.*

I state that x, y, and z can again be expanded in powers of the *four* variables μ, x_0, y_0, and z_0 however large t might be provided that the four variables μ, x_0, y_0, and z_0 are sufficiently small. There is just the same a special case to which I will return.

In fact, we first find three series

$$x = \varphi(t, \mu, x_0, y_0, z_0), \quad y = \varphi_1(t, \mu, x_0, y_0, z_0), \quad z = \varphi_2(t, \mu, x_0, y_0, z_0)$$

which define x, y, and z for sufficiently small values of μ, x_0, y_0, and z_0 and when

$$|t| < \rho,$$

where ρ is the radius of convergence of the series. If therefore t_1 is a point inside the circle of convergence and if x_1, y_1, and z_1 are values of x, y, and z for $t = t_1$, it can be seen that x_1, y_1, and z_1 are analytic functions of μ, x_0, y_0, and z_0, which means they are expandable in powers of these variables if they are sufficiently small.

Next let x_1^0, y_1^0, and z_1^0 be the values of x_1, y_1, and z_1 for

$$\mu = x_0 = y_0 = z_0 = 0.$$

Having set that, then in the neighborhood of the point $t = t_1$

$$\begin{aligned} x &= \varphi'\left(t - t_1, \mu, x_1 - x_1^0, y_1 - y_1^0, z_1 - z_1^0\right), \\ y &= \varphi_1'\left(t - t_1, \mu, x_1 - x_1^0, y_1 - y_1^0, z_1 - z_1^0\right), \\ z &= \varphi_2'\left(t - t_1, \mu, x_1 - x_1^0, y_1 - y_1^0, z_1 - z_1^0\right). \end{aligned} \quad (3)$$

The series φ', φ_1', and φ_2', which are completely analogous to the series φ, φ_1, and φ_2, are defined as follows.

They satisfy the differential equations; they are expanded in powers of $t - t_1$, μ, $x_1 - x_1^0$, $y_1 - y_1^0$, and $z_1 - z_1^0$; they reduce to x_1, y_1, and z_1 for $t = t_1$.

They converge if $t - t_1$, μ, $x_1 - x_1^0$, $y_1 - y_1^0$, and $z_1 - z_1^0$ are small enough and if

$$|t - t_1| < \rho_1$$

where ρ_1 is the radius of the new circle of convergence C_1.

If t is a point inside this new circle of convergence C_1, then it can be seen that x, y, and z will be mapping functions of μ, $x_1 - x_1^0$, $y_1 - y_1^0$, and $z_1 - z_1^0$. But $x_1 - x_1^0$,

2 Calculation of Limits

$y_1 - y_1^0$, and $z_1 - z_1^0$ are already analytic functions of μ, x_0, y_0, and z_0. Therefore, for any point t inside the circle

C_1, the three quantities x, y, and z are mapping functions of μ, x_0, y_0, and z_0 and are expandable in powers of these variables if they are sufficiently small.

Now assume that the point t is outside of the circle C_1; the theorem will still be true. It is clear in fact that to prove it for an arbitrary value of t it is sufficient to repeat the previous reasoning a sufficient number of times, provided that the radii $\rho_1, \rho_2, \ldots$ of the circles of convergence considered successively remain greater than a given quantity.

Additionally this convergence will be uniform for any value of t less than t_0, however large t_0 is.

One would only be stopped in one case.

The Cauchy theorem stops being true if the functions f_1 and f_2 are no longer mappings of x, y, and z; for example if they become infinite or stop being one-to-one.

If the functions f, f_1, and f_2 cannot be expanded in increasing powers of μ, $x - x_1^0$, $y - y_1^0$, and $z - z_1^0$, there will in general not exist three series φ', φ_1', and φ_2' of the form (3) satisfying the differential equations.

It is then said that the point

$$x = x_1^0, \quad y = y_1^0, \quad z = z_1^0,$$

is a singular point.

If therefore, by varying t, one saw the moving point (x, y, z) pass by a singular point, our theorem would then be wrong. If with t varying from $t = 0$ to $t = t_0$, the moving point (x, y, z) does not pass by any singular point, the radius of convergence of the Cauchy series cannot become zero and it is possible to assign it a lower bound such that the three functions x, y, and z could be expanded in powers of μ, x, y, and z for any value of t less than t_0. But if for $t = t_0$, the point is coincident with a singular point, the theorem stops being true for values of t greater than t_0.

Our theorem therefore comprises one exceptional case. But this case will not come up in the three-body problem and we do not need to be concerned about it. Let in fact:

$$(x_1, y_1, z_1), \quad (x_2, y_2, z_2), \quad (x_3, y_3, z_3)$$

be the coordinates of three bodies; let r_{23}, r_{13}, and r_{12} be their mutual distances; and let m_1, m_2, and m_3 be their masses. The equations for the problem will then have the following form:

$$\frac{d^2 x_1}{dt^2} = \frac{m_2(x_2 - x_1)}{r_{12}^3} + \frac{m_3(x_3 - x_1)}{r_{13}^3}.$$

The right-hand side of this equation could stop being analytic in x_1, y_1, z_1, x_2, y_2, z_2, x_3, y_3, z_3 only if one of the three distances r_{23}, r_{13}, and r_{12} became zero; meaning if the two bodies collided. Therefore, we will never apply our theorem unless we are certain that such a collision could not occur.

The same result could also be established in another way. Take the equations:

$$\frac{dx}{dt} = f(x, y, z, \mu),$$
$$\frac{dy}{dt} = f_1(x, y, z, \mu),$$
$$\frac{dz}{dt} = f_2(x, y, z, \mu).$$

The functions f, f_1, and f_2 could in general be expanded in increasing powers of $x - x_0$, $y - y_0$, $z - z_0$, and $\mu - \mu_0$ for values of x, y, z, and μ sufficiently close to x_0, y_0, z_0, and μ_0. If there is a set of values x_0, y_0, z_0, and μ_0 for which this is not the case, I will state that this set of values is *one of the singular points* of our differential equations.

Having said that, these equations will allow a solution such that x, y, and z are zero when t is zero; and this solution will obviously depend on μ. Let:

$$x = \omega_1(t - \mu), \quad y = \omega_2(t - \mu), \quad x = \omega_3(t - \mu)$$

be this solution. It follows from the definition of this solution itself that, whatever μ is:

$$\omega_1(0, \mu) = \omega_2(0, \mu) = \omega_3(0, \mu) = 0.$$

In most applications, the integration could be done for $\mu = 0$, such that the functions $\omega_1(t, 0)$, $\omega_2(t, 0)$, and $\omega_3(t, 0)$ will be known. I assume that the set of values

$$\omega_1(t, 0), \quad \omega_2(t, 0), \quad \omega_3(t, 0), \quad 0$$

is not a singular point of our differential equations for any of the values of t included between 0 and t_1.

To state it more conveniently but less correctly, I will say that the specific solution

$$x = \omega_1(t, 0), \quad y = \omega_2(t, 0), \quad z = \omega_3(t, 0),$$

does not pass through any singular point.

If that were not the case, we would find ourselves in the special case which I spoke about above.

If, in contrast, that is the case, which is what I will assume, then I state that the expressions $\omega_1(t_1, \mu)$, $\omega_2(t_1, \mu)$, and $\omega_3(t_1, \mu)$ are functions of μ which are expandable in increasing powers of this variable.

2 Calculation of Limits

In fact, set

$$x = \xi + \omega_1(t,0), \quad y = \eta + \omega_2(t,0), \quad z = \eta + \omega_3(t,0),$$

and the differential equations will become:

$$\begin{aligned}\frac{d\xi}{dt} &= \varphi(\xi,\eta,\zeta,t,\mu),\\ \frac{d\eta}{dt} &= \varphi_1(\xi,\eta,\zeta,t,\mu),\\ \frac{d\zeta}{dt} &= \varphi_2(\xi,\eta,\zeta,t,\mu).\end{aligned} \quad (4)$$

It follows from this assumption that we have made that for all values of t included between 0 and t_1, the functions φ, φ_1, and φ_2 are expandable in powers of ξ, η, ζ, and μ with the expansion coefficients being functions of time.

I observe that for $\mu = 0$, the differential equations must be satisfied for

$$\xi = \eta = \zeta = 0,$$

which means that φ, φ_1, and φ_2 become zero when μ, ξ, η, and ζ also become zero simultaneously.

It is then possible to find two positive numbers M and α such that, for all values of t included between 0 and t_1, each coefficient of the expansion of φ, φ_1, or φ_2 according to increasing powers of ξ, η, ζ, and μ is smaller in absolute value than the corresponding coefficient of the expansion of:

$$\frac{M(\xi+\eta+\zeta+\mu)}{1-\alpha(\xi+\eta+\zeta+\mu)}$$

or more especially than the corresponding coefficient of the expansion of:

$$\psi(\xi,\eta,\zeta,\mu) = \frac{M(\xi+\eta+\zeta+\mu)[1+\alpha(\xi+\eta+\zeta+\mu)]}{1-\alpha(\xi+\eta+\zeta+\mu)}.$$

We therefore compare Eq. (4) to the following:

$$\frac{d\xi}{dt} = \frac{d\eta}{dt} = \frac{d\zeta}{dt} = \psi(\xi,\eta,\zeta,\mu). \quad (5)$$

The solution of Eq. (4), which is such that ξ, η, and ζ become zero together for $t = 0$, is written:

$$\xi = \omega_1(t,\mu) - \omega_1(t,0), \quad \eta = \omega_2(t,\mu) - \omega_2(t,0), \quad \zeta = \omega_3(t,\mu) - \omega_3(t,0).$$

From another perspective Eq. (5) allows a solution:

$$\xi = \eta = \zeta = \omega'(t,\mu)$$

such that ξ, η, and ζ are zero when t is.

By reasoning as Cauchy did, it can be seen that if $\omega'(t,\mu)$ is expandable in increasing powers of μ, it must be the same for $\omega_1(t,\mu) - \omega_1(t,0)$, $\omega_2(t,\mu) - \omega_2(t,0)$, and $\omega_3(t,\mu) - \omega_3(t,0)$ and that each coefficient of the expansion of these last three functions is smaller in absolute value than the corresponding coefficient of $\omega'(t,\mu)$ at least for all values of t such that

$$0 < t < t_1.$$

Hence Eq. (5) is easy to integrate and it can easily be verified that $\omega'(t,\mu)$ is expandable in powers of μ. Therefore ξ, η, and ζ can also be expanded in powers of μ provided that

$$0 < t < t_1.$$

Which was to be proved.

Theorem III *With that established, let*:

$$x = \omega_1(t,\mu,x_0,y_0,z_0), \quad y = \omega_2(t,\mu,x_0,y_0,z_0), \quad z = \omega_3(t,\mu,x_0,y_0,z_0)$$

be that of the solutions of our differential equations, which for $t = 0$ is such that

$$x = x_0, \quad y = y_0, \quad z = z_0.$$

Consider the functions:

$$\omega_1(t_1+\tau,\mu,x_0,y_0,z_0), \quad \omega_2(t_1+\tau,\mu,x_0,y_0,z_0), \quad \omega_3(t_1+\tau,\mu,x_0,y_0,z_0).$$

I state that they are expandable in powers of μ, x_0, y_0, z_0, and τ provided that these quantities are sufficiently small.

In fact set:

$$x = x' + x_0, \quad y = y' + y_0, \quad z = z' + z_0, \quad t = t'\frac{t_1+\tau}{t_1}.$$

Our equations will become:

$$\frac{dx'}{dt} = \left(1 + \frac{\tau}{t_1}\right) f(x'+x_0, y'+y_0, z'+z_0, \mu),$$

$$\frac{dy'}{dt} = \left(1 + \frac{\tau}{t_1}\right) f_1(x'+x_0, y'+y_0, z'+z_0, \mu),$$

$$\frac{dz'}{dt} = \left(1 + \frac{\tau}{t_1}\right) f_2(x'+x_0, y'+y_0, z'+z_0, \mu).$$

2 Calculation of Limits

These equations contain five arbitrary parameters specifically

$$\mu, x_0, y_0, z_0, \tau.$$

Consider the solution of these equations which is such that x', y', and z' are zero when t' is; let:

$$\begin{aligned} x' &= \omega'_1(t', \mu, x_0, y_0, z_0, \tau), \\ y' &= \omega'_2(t', \mu, x_0, y_0, z_0, \tau), \\ z' &= \omega'_3(t', \mu, x_0, y_0, z_0, \tau). \end{aligned}$$

It follows from what we just saw that if we set $t' = t_1$, the expressions:

$$\begin{aligned} \omega'_1(t_1, \mu, x_0, y_0, z_0, \tau), \\ \omega'_2(t_1, \mu, x_0, y_0, z_0, \tau), \\ \omega'_3(t_1, \mu, x_0, y_0, z_0, \tau). \end{aligned}$$

are expandable in powers of μ, x_0, y_0, z_0, and τ. But it is obvious that we have:

$$\begin{aligned} \omega'_1(t_1, \mu, x_0, y_0, z_0, \tau) &= \omega_1(t_1 + \tau, \mu, x_0, y_0, z_0), \\ \omega'_2(t_1, \mu, x_0, y_0, z_0, \tau) &= \omega_2(t_1 + \tau, \mu, x_0, y_0, z_0), \\ \omega'_3(t_1, \mu, x_0, y_0, z_0, \tau) &= \omega_3(t_1 + \tau, \mu, x_0, y_0, z_0). \end{aligned} \quad (6)$$

Therefore the right-hand side of Eq. (6) can also be expanded in powers of μ, x_0, y_0, z_0, and τ.
Which was to be proved.

Theorem IV Cauchy *drew another extremely important theorem from the calculation of limits.*

This is what that theorem is:

If there are $n+p$ quantities $y_1, y_2, \ldots y_n, x_1, x_2, \ldots x_p$ between which there are n relations:

$$\begin{aligned} f_1(y_1, y_2, \ldots y_n, x_1, x_2, \ldots x_p) &= 0, \\ f_2(y_1, y_2, \ldots y_n, x_1, x_2, \ldots x_p) &= 0, \\ &\vdots \\ f_n(y_1, y_2, \ldots y_n, x_1, x_2, \ldots x_p) &= 0; \end{aligned} \quad (7)$$

if the f are expandable in powers of x and y and are zero when these $n+p$ variables are zero;

if, finally, the Jacobian determinant of these f with respect to y is not zero when the x and y are both zero;

then the n unknowns y can be determined from Eq. (7) in the form of series expanded in increasing powers of $x_1, x_2, \ldots x_p$.

In fact consider x_1 as the only independent variable and $x_2, x_3, \ldots x_p$ as arbitrary parameters, we can then replace Eq. (7) by the n differential equations:

$$\frac{df_i}{dy_1}\frac{dy_1}{dx_1} + \frac{df_i}{dy_2}\frac{dy_2}{dx_1} + \cdots + \frac{df_i}{dy_n}\frac{dy_n}{dx_1} + \frac{df_i}{dx_1} = 0. \quad \text{(for } i = 1, 2, \ldots n\text{)} \tag{8}$$

This brings us back to the case that we were just looking at.

In particular if $f(y, x_1, x_2, \ldots x_n)$ is a function which is expandable in powers of $y, x_1, x_2, \ldots x_n$; if when all x and y are zero at the same time then:

$$f = 0, \quad \frac{df}{dy} \neq 0;$$

if finally y is defined by the equality

$$f = 0,$$

y is then expandable in powers of x.

All that is left is for us to examine what happens when the Jacobian determinant of f with respect to y is zero. This question has been the subject of extensive research which I will not go into here, but it is useful to cite the work of Mr. Puiseux on the roots of algebraic equations as the leader among the work. I too had the opportunity to be involved in analogous research in the first part of my inaugural thesis (Paris, Gauthier-Villars, 1879). I will therefore limit myself to stating the following theorems while limiting myself to referring to conventional treatises or my thesis for the proofs.

Theorem V *Let y be a function of x defined by the equation*

$$f(y, x) = 0 \tag{9}$$

where f is expandable in powers of x and y.

I assume that for $x = y = 0$, then f is zero as well as:

$$\frac{df}{dy}, \frac{d^2f}{dy^2}, \ldots \frac{d^{m-1}f}{dy^{m-1}},$$

but that $d^m f/dy^m$ does not become zero.

There will exist m series of the following form which will satisfy Eq. (9):

$$y = a_1 x^{1/n} + a_2 x^{2/n} + a_3 x^{3/n} + \cdots \tag{10}$$

(where n is a positive integer and where $a_1, a_2, \ldots$ are constant coefficients).

Corollary I *If the series* (10) *satisfies Eq.* (9), *the same is true of the series*:

$$y = a_1 \alpha x^{1/n} + a_2 \alpha^2 x^{2/n} + a_3 \alpha^3 x^{3/n} + \cdots$$

where α is an nth root of one.

Corollary II *The number of series of the form* (10) *expanded in powers of $x^{1/n}$, (which cannot be expanded in powers of $x^{1/p}, p < n$) is divisible by n.*

Corollary III *If $k_1 n_1$ is the number of series* (10) *which are expandable in powers of x^{1/n_1}, if $k_2 n_2$ is the number of series* (10) *which are expandable in powers of x^{1/n_2}, ..., if $k_p n_p$ is the number of series* (10) *which are expandable in powers of x^{1/n_p}, then it follows*:

$$k_1 n_1 + k_2 n_2 + \cdots k_p n_p = m$$

Hence it can be concluded that if m is odd, at least one of the numbers $n_1, n_2, \ldots n_p$ is also odd.

Theorem VI *If there are equations*

$$\begin{aligned} f_1(y_1, y_2, \ldots y_p, x) &= 0, \\ f_2(y_1, y_2, \ldots y_p, x) &= 0, \\ &\vdots \\ f_p(y_1, y_2, \ldots y_p, x) &= 0, \end{aligned} \qquad (11)$$

whose left-hand sides are expandable in powers of y and x and are zero when these variables are zero, then

$$y_2, y_3, \ldots y_p$$

can always be eliminated between these equations and a single equation

$$f(y_1, x) = 0$$

of the same form as Eq. (9) *of the previous theorem can be obtained.*

There would only be an exception if Eq. (11) was no longer distinct.

Corollary to theorems V and VI Theorem IV applies every time that the Jacobian determinant of f is not zero; meaning every time that the equations

$$f_1 = f_2 = \cdots f_n = 0 \qquad (7)$$

allows
$$y_1 = y_2 = \cdots y_n = 0$$

as a *single* solution when the *x* are all zero.

It follows from theorems V and VI and their corollaries stated above that the theorem IV are still true if there are multiple solutions, *provided that the order of multiplicity is odd.*

3 Application of the Calculation of Limits to Partial Differential Equations

Cauchy already applied the method of calculation of limits to partial differential equations. Ms. Kowalevski considerably simplified Cauchy's proof and gave the theorem its final form.

Here is what Ms. Kowalevski's theorem (Crelle's Journal, volume 80) consists of.

Consider a family of partial differential equations defining n unknowns $z_1, z_2, \ldots z_n$ as functions of p independent variables.

Suppose that this system is written:

$$\begin{aligned}\frac{dz_1}{dx_1} &= f_1\left(x_1, x_2, \ldots x_p; \frac{dz_i}{dx_2}, \frac{dz_i}{dx_3}, \ldots \frac{dz_i}{dx_p}\right), \\ \frac{dz_2}{dx_1} &= f_2\left(x_1, x_2, \ldots x_p; \frac{dz_i}{dx_2}, \frac{dz_i}{dx_3}, \ldots \frac{dz_i}{dx_p}\right), \\ &\vdots \\ \frac{dz_n}{dx_1} &= f_n\left(x_1, x_2, \ldots x_p; \frac{dz_i}{dx_2}, \frac{dz_i}{dx_3}, \ldots \frac{dz_i}{dx_p}\right),\end{aligned} \quad (1)$$

where $f_1, f_2, \ldots f_n$ are expanded in powers of

$$x_1, x_2, \ldots x_p$$

and of

$$\frac{dz_i}{dx_k} - \alpha_{ik}$$

(*i* takes the values $1, 2, \ldots n$; *k* takes the values $2, 3, \ldots p$; finally the α_{ik} are arbitrary constants).

Now let:

$$\psi_1(x_2, x_3, \ldots x_p), \quad \psi_2(x_2, x_3, \ldots x_p), \quad \ldots \quad \psi_n(x_2, x_3, \ldots x_p),$$

be *n arbitrary* given functions, expanded in increasing powers of $x_2, x_3, \ldots x_p$ and such that:

$$\frac{d\psi_i}{dx_k} = \alpha_{ik}$$

for

$$x_2 = x_3 = \cdots x_p = 0.$$

There will exist n functions

$$z_1 = \varphi_1(x_1, x_2, \ldots x_p), \quad z_2 = \varphi_2(x_1, x_2, \ldots x_p), \quad \ldots \quad z_n = \varphi_n(x_1, x_2, \ldots x_p)$$

which are expandable in powers of $x_1, x_2, \ldots x_p$, which will satisfy Eq. (1) and which will reduce, respectively, to $\psi_1, \psi_2, \ldots \psi_n$ for $x_1 = 0$.

I have myself sought to extend the results obtained by Ms. Kowalevski (*Inaugural Thesis*, Paris, Gauthier-Villars, 1879) and I have studied in detail the cases that the renowned mathematician had set aside.

In particular I focused on the equation:

$$X_1 \frac{dz}{dx_1} + X_2 \frac{dz}{dx_2} + \cdots X_n \frac{dz}{dx_n} = \lambda_1 z, \qquad (2)$$

where $X_1, X_2, \ldots X_n$ are expandable in powers of $x_1, x_2, \ldots x_n$; I additionally suppose that in the expansion of $X_1, X_2, \ldots X_n$, there are no constant terms and that the first-degree terms reduce, respectively, to $\lambda_1 x_1, \lambda_2 x_2, \ldots \lambda_n x_n$, such that

$$X_i = \lambda_i x_i - Y_i,$$

where Y_i designates a series of terms of at least second-order in $x_1, x_2, \ldots x_n$.

I proved that for some conditions this equation allows a mapping integral which is expandable in powers of $x_1, x_2, \ldots x_n$.

For this integral to exist, it is sufficient that:

(1) the convex polygon which contains the n points $\lambda_1, \lambda_2, \ldots \lambda_n$ does not contain the origin,
(2) there is no relation of the form

$$m_2 \lambda_2 + \cdots m_n \lambda_n = \lambda_1$$

where the m are positive integers whose sum is greater than one.[1]

I am seeking to generalize the results obtained in my thesis. Instead of Eq. (2) consider the following equation:

$$\frac{dz}{dt} + X_1 \frac{dz}{dx_1} + X_2 \frac{dz}{dx_2} + \cdots X_n \frac{dz}{dx_n} = \lambda_1 z. \qquad (3)$$

Again, we have

$$X_i = \lambda_i x_i - Y_i,$$

where Y_i designates a function expanded in powers of $x_1, x_2, \ldots x_n$ and includes only terms of at least second-order in these n variables. But Y_i does not depend only on x, it also depends on t, such that the coefficients of the expansion of Y_i in powers of x are functions of t. We will assume that these are periodic functions of t with period 2π expanded in sines and cosines of multiples of t.

I propose to look for the cases where Eq. (3) has a mapping integral expanded in powers of $x_1, x_2, \ldots x_n$ and such that the coefficients of the expansion are periodic functions of t.

First look at what the form of Y_i is going to be. We are going to expand Y_i in increasing powers of $x_1, x_2, \ldots x_n$; let us consider the term in

$$x_1^{\alpha_1} x_2^{\alpha_2} \ldots x_n^{\alpha_n}.$$

Since the coefficient of this term is a periodic function of t, the coefficient is expandable in sines and cosines of multiples of t, or, what amounts to the same thing, positive and negative powers of $e^{t\sqrt{-1}}$.

We will therefore write:

$$Y_i = \sum C_{i,\beta,\alpha_1,\alpha_2,\ldots\alpha_n} e^{\beta t \sqrt{-1}} x_1^{\alpha_1} x_2^{\alpha_2} \ldots x_n^{\alpha_n}.$$

where the C are constant coefficients; β is a positive or negative integer; $\alpha_1, \alpha_2, \ldots \alpha_n$ are positive integers such that

$$\alpha_1 + \alpha_2 + \cdots \alpha_n \geq 2.$$

---

[1] In my thesis, I did not state this restriction and I did not assume that the sum of the m was larger than 1. It would therefore seem that the theorem is incorrect when for example $\lambda_2 = \lambda_1$. This is not the case. If we had

$$m_2 \lambda_2 + \cdots m_n \lambda_n = \lambda_1 \text{ for } (m_2 + m_3 + \cdots m_n > 1)$$

some coefficients of the expansion would take the form $A/0$ and would become infinite. This is the reason why we had to assume that such a relationship does not hold. If on the other hand $\lambda_2 = \lambda_1$, then some coefficients would take the form $0/0$.

3 Application of the Calculation of Limits to Partial Differential Equations

Sometimes, I will also write it leaving off the indices:

$$Y_i = \sum C e^{\beta t \sqrt{-1}} x_1^{\alpha_1} x_2^{\alpha_2} \cdots x_n^{\alpha_n}.$$

Now set:

$$Y_i' = \sum |C| e^{\beta t \sqrt{-1}} x_1^{\alpha_1} x_2^{\alpha_2} \cdots x_n^{\alpha_n}$$

and consider the following equation:

$$(\lambda_1' x_1 - Y_1') \frac{dz}{dx_1} + (\lambda_2' x_2 - Y_2') \frac{dz}{dx_2} + \cdots (\lambda_n' x_n - Y_n') \frac{dz}{dx_n} = \lambda_1' z. \quad (4)$$

In this equation, dz/dt no longer appears; we can therefore regard t as an arbitrary parameter and $x_1, x_2, \ldots x_n$ as the only independent variables. If therefore the quantities $\lambda_1', \lambda_2', \ldots \lambda_n'$ satisfy the conditions that we stated above, Eq. (4) [which has the same form as Eq. (2)] will allow a mapping integral.

We will assume:

$$\lambda_1' = \lambda_2' = \cdots \lambda_n'.$$

We will additionally assume that λ_1' is real and positive.
Having assumed that, let:

$$z = \sum A_{\beta, \alpha_1, \alpha_2, \ldots \alpha_n} e^{\beta t \sqrt{-1}} x_1^{\alpha_1} x_2^{\alpha_2} \ldots x_n^{\alpha_n} \quad (5)$$

be a series formally satisfying Eq. (3). How would it be possible to calculate the coefficients A by iteration?

By writing Eq. (3) in the form:

$$\frac{dz}{dt} + \lambda_1 x_1 \frac{dz}{dx_1} + \cdots \lambda_n x_n \frac{dz}{dx_n} - \lambda_1 z = Y_1 \frac{dz}{dx_1} + Y_2 \frac{dz}{dx_2} + \cdots Y_n \frac{dz}{dx_n}$$

and by aligning the left- and right-sides, we find:

$$A_{\beta, \alpha_1, \alpha_2, \cdots \alpha_n} \left[\beta \sqrt{-1} + \lambda_1 \alpha_1 + \lambda_2 \alpha_2 + \cdots \lambda_n \alpha_n - \lambda_1 \right] = P[C, A].$$

where $P[C, A]$ is an integer polynomial with positive coefficients in C and the coefficients A already calculated.

Now let

$$z = \sum A'_{\beta, \alpha_1, \alpha_2, \cdots \alpha_n} e^{\beta t \sqrt{-1}} x_1^{\alpha_1} x_2^{\alpha_2} \cdots x_n^{\alpha_n} \quad (6)$$

be a series satisfying Eq. (4). To calculate the coefficients A' we will write Eq. (4) in the form:

$$\lambda'_1 x_1 \frac{dz}{dx_1} + \lambda'_2 x_2 \frac{dz}{dx_2} + \cdots \lambda'_n x_n \frac{dz}{dx_n} - \lambda'_1 z = Y'_1 \frac{dz}{dx_1} + Y'_2 \frac{dz}{dx_2} + \cdots Y'_n \frac{dz}{dx_n}.$$

By aligning the left- and right-sides, we will find:

$$A'_{\beta,\alpha_1,\alpha_2,\cdots\alpha_n} [\lambda'_1 \alpha_1 + \lambda'_2 \alpha_2 + \cdots \lambda'_n \alpha_n - \lambda'_1] = P[|C|, A'].$$

$P[|C|, A']$ is only different from $P[C, A]$ because the C are replaced by their magnitudes and the A by the A'.

Since the λ' are positive real numbers as are the coefficients of the polynomial P, the A' will also be real and positive.

So that we next have:

$$|A_{\beta,\alpha_1,\alpha_2,\cdots\alpha_n}| < A'_{\beta,\alpha_1,\alpha_2,\ldots\alpha_n},$$

it is sufficient that we always have:

$$\lambda'_1 \alpha_1 + \lambda'_2 \alpha_2 + \cdots \lambda'_n \alpha_n - \lambda'_1 < \left| \beta\sqrt{-1} + \lambda_1 \alpha_1 + \lambda_2 \alpha_2 + \cdots \lambda_n \alpha_n - \lambda_1 \right|$$

or

$$\lambda'_1 < \left| \frac{\beta\sqrt{-1} + \lambda_1(\alpha_1 - 1) + \lambda_2 \alpha_2 + \cdots \lambda_n \alpha_n}{(\alpha_1 - 1) + \alpha_2 + \cdots \alpha_n} \right|. \quad (7)$$

If λ'_1 is chosen so as to satisfy the inequality (7),

$$|A| < A'.$$

Hence the series (6) converges, and therefore the series (5) will likewise converge.

And so for the series (5) to converge, it is sufficient to be able to find a positive quantity λ'_1 satisfying the inequality (7) for all positive integer values of α, and for all positive and negative integer values of β.

Start by noting the right-hand side of the inequality (7) is always greater than:

$$\left| \frac{\beta\sqrt{-1} + \lambda_1(\alpha_1 - 1) + \lambda_2 \alpha_2 + \cdots \lambda_n \alpha_n}{|\beta| + (\alpha_1 - 1) + \alpha_2 + \cdots \alpha_n} \right|. \quad (8)$$

3 Application of the Calculation of Limits to Partial Differential Equations

It will therefore be sufficient that λ'_1 be smaller than expression (8). Now, this expression (8) is the magnitude of a certain imaginary quantity represented by a certain point G. Now, it is easy to see that this point G is nothing other than the center of gravity of the following $n+2$ masses:

(1) n masses equal, respectively, to $\alpha_1, \alpha_2, \ldots \alpha_n$ and located, respectively, at points $\lambda_1, \lambda_2, \ldots \lambda_n$;
(2) a mass equal to β and located at the point $+\sqrt{-1}$ or at the point $-\sqrt{-1}$;
(3) a mass equal to -1 located at point λ_1.

All these masses are positive except for the last one.

The condition needs to be found so that the distance OG is always greater than some bound λ'_1.

First combine the first $n+1$ masses; the result is a mass:

$$M = \alpha_1 + \alpha_2 + \cdots \alpha_n + |\beta|$$

located at some point G' and since these first $n+1$ masses are positive, the point G' will be located inside one or the other of two surrounding convex polygons, the first with the $n+1$ points

$$\lambda_1, \lambda_2, \ldots \lambda_n \quad \text{and} \quad +\sqrt{-1}$$

and the second with the $n+1$ points

$$\lambda_1, \lambda_2, \ldots \lambda_n \quad \text{and} \quad -\sqrt{-1}.$$

If neither of these convex polygons contains the origin, then the distance OG' can be assigned a lower bound μ and one can write:

$$OG' > \mu.$$

It remains to combine the mass M located at G' and the mass -1 located at λ_1. The result will then be a mass $M - 1$ located at G. We will then obviously have:

$$OG > OG' - OG',$$

$$GG' = \frac{G'\lambda_1}{M-1} < \frac{OG'}{M-1} + \frac{O\lambda_1}{M-1},$$

hence

$$OG > OG'\frac{M-2}{M-1} - \frac{O\lambda_1}{M-1} > \mu\frac{M-2}{M-1} - \frac{O\lambda_1}{M-1}.$$

If therefore:

$$M > \frac{3\mu + 20\lambda_1}{\mu},$$

then the inequality

$$OG > \frac{\mu}{2} \tag{9}$$

will be satisfied.

Therefore there are only a finite number of combinations of integer numbers:

$$\alpha_1, \alpha_2, \cdots \alpha_n, \beta$$

for which the inequality (9) might not be satisfied.

If OG is zero for none of these combinations, we are certain to be able to assign a lower bound λ_1 to OG.

We are therefore led to state the following rule:

For Eq. (3) to allow an integral which is expandable in powers of x and which is periodic in t, it is sufficient that:

(1) neither of the two circumscribed convex polygons—the first with points $\lambda_1, \lambda_2, \ldots \lambda_n$ and $+\sqrt{-1}$ and the second with points $\lambda_1, \lambda_2, \ldots \lambda_n$ and $-\sqrt{-1}$—contain the origin,
(2) there is no relation between the quantities of the form

$$\beta\sqrt{-1} + \lambda_1\alpha_1 + \lambda_2\alpha_2 + \cdots \lambda_n\alpha_n = \lambda_1,$$

where the α are positive integers and β is a positive or negative integer.

So that is a generalization of the theorem which was proved in my thesis. Now, a number of consequences follow from this theorem. Let us see if it will be possible to draw similar ones from the generalized theorem.

For that we are absolutely going to follow the same path as in the cited thesis. Consider the equation:

$$\frac{dz}{dt} + X_1 \frac{dz}{dx_1} + X_2 \frac{dz}{dx_2} + \cdots X_n \frac{dz}{dx_n} = 0, \tag{10}$$

Obtained by dropping the right-hand side of Eq. (3).
Further, consider the equation:

$$\frac{dz}{dt} + X_1 \frac{dz}{dx_1} + X_2 \frac{dz}{dx_2} + \cdots X_n \frac{dz}{dx_n} = \lambda_1 z \tag{3}$$

3 Application of the Calculation of Limits to Partial Differential Equations

and the equation:

$$\frac{dz}{dt} + X_1 \frac{dz}{dx_1} + X_2 \frac{dz}{dx_2} + \ldots X_n \frac{dz}{dx_n} = \lambda_2 z. \tag{11}$$

If the λ satisfy the conditions which we just stated, Eq. (3) will then allow an integral

$$z = T_1$$

where T_1 is ordered by powers of x and periodic in t.

Similarly Eq. (11) will allow an integral

$$z = T_2$$

where T_2 has the same form as T_1.

From this, we conclude that Eq. (10) allows as a particular integral:

$$T_1^{\frac{1}{\lambda_1}} T_2^{-\frac{1}{\lambda_2}}.$$

Since we can successively replace $\lambda_1 z$ by $\lambda_2 z, \lambda_3 z, \ldots \lambda_n z$ in the left-hand side of (3) and we thereby get $n-1$ equations analogous to Eq. (11), we can then conclude that Eq. (10) allows $n-1$ specific integrals:

$$T_1^{\frac{1}{\lambda_1}} T_2^{-\frac{1}{\lambda_2}}, T_1^{\frac{1}{\lambda_1}} T_3^{-\frac{1}{\lambda_3}}, \ldots T_1^{\frac{1}{\lambda_1}} T_n^{-\frac{1}{\lambda_n}}$$

where $T_2, T_3, \ldots T_n$ have the same form as T_1.

To get a general integral of (10), an nth specific integral is still needed. To get that, consider the equation:

$$\frac{dz}{dt} + X_1 \frac{dz}{dx_1} + X_2 \frac{dz}{dx_2} + \cdots X_n \frac{dz}{dx_n} = z. \tag{12}$$

This equation will allow $z = e^t$ as a specific integral.

From this we conclude that Eq. (10) allows the following as specific integrals:

$$T_1 e^{-\lambda_1 t}, T_2 e^{-\lambda_2 t}, \ldots T_n e^{-\lambda_n t},$$

such that the general integral of this Eq. (10) will be:

$$z = \text{arbitrary function of } \left(T_1 e^{-\lambda_1 t}, T_2 e^{-\lambda_2 t}, \cdots T_n e^{-\lambda_n t} \right)$$

In other words, the differential equations:

$$dt = \frac{dx_1}{X_1} = \frac{dx_2}{X_2} = \cdots \frac{dx_n}{X_n} \tag{10'}$$

will allow

$$T_1 = K_1 e^{\lambda_1 t}, \quad T_2 = K_2 e^{\lambda_2 t}, \quad \ldots \quad T_n = K_n e^{\lambda_n t},$$

as general integrals, where $K_1, K_2, \ldots K_n$ are n constants of integration.

This theorem can be regarded as the generalization of the one that I proved on page 70 of my thesis.

Now suppose that we are looking to determine the first $p\ x$ variables, specifically:

$$x_1, x_2, \ldots x_p$$

as functions of the $n - p$ others, specifically

$$x_{p+1}, x_{p+2}, \ldots x_n$$

and of t, using the following equations:

$$\frac{dx_1}{dt} + X_{p+1}\frac{dx_1}{dx_{p+1}} + X_{p+2}\frac{dx_1}{dx_{p+2}} + \cdots X_n\frac{dx_1}{dx_n} = X_1,$$
$$\frac{dx_2}{dt} + X_{p+1}\frac{dx_2}{dx_{p+1}} + X_{p+2}\frac{dx_2}{dx_{p+2}} + \cdots X_n\frac{dx_2}{dx_n} = X_2, \tag{13}$$
$$\vdots$$
$$\frac{dx_p}{dt} + X_{p+1}\frac{dx_p}{dx_{p+1}} + X_{p+2}\frac{dx_p}{dx_{p+2}} + \cdots X_n\frac{dx_p}{dx_n} = X_p.$$

It is easy to see that the general integral of Eq. (13) can be written as:

$$\varphi_1 = \varphi_2 = \cdots \varphi_p = 0 \tag{14}$$

where $\varphi_1, \varphi_2, \ldots \varphi_p$ represent p arbitrary functions of

$$T_1 e^{-\lambda_1 t}, T_2 e^{-\lambda_2 t}, \ldots T_n e^{-\lambda_n t}.$$

Specifically, we let:

$$\varphi_1 = T_1 e^{-\lambda_1 t}, \quad \varphi_2 = T_2 e^{-\lambda_2 t}, \quad \cdots \quad \varphi_p = T_p e^{-\lambda_p t}.$$

Equation (14) can then be written:

$$T_1 = T_2 = \cdots T_p = 0. \tag{14'}$$

3 Application of the Calculation of Limits to Partial Differential Equations   27

From Eq. (14'), the $x_1, x_2, \ldots x_p$ can be drawn as functions of $x_{p+1}, x_{p+2}, \ldots x_n$ and t and it will be seen that these are mapping functions of $x_{p+1}, x_{p+2}, \ldots x_n$ and periodic functions of t.

Therefore Eq. (13) allows a solution expandable in increasing powers of $x_{p+1}, x_{p+2}, \ldots x_n$ and in sines and cosines of multiples of t.

This theorem is proved when the λ satisfy the conditions stated above; let us see how it can be extended to cases where these conditions are not satisfied. For that, I will follow the same process as in the fourth part of my research on *Courbes définies par les équations différentielles (Curves Defined by Differential Equations;* Journal de Liouville, fourth series, volume 2, pages 156–157).

Let us propose to calculate the coefficients of the mapping integral for Eq. (13; assuming that this integral exists) and for that purpose let us write this Eq. (13) in the following form:

$$\frac{dx_i}{dt} + \lambda_{p+1}x_{p+1}\frac{dx_i}{dx_{p+1}} + \lambda_{p+2}x_{p+2}\frac{dx_i}{dx_{p+2}} + \cdots \lambda_n x_n \frac{dx_i}{dx_n} - \lambda_i x_i$$
$$= Y_{p+1}\frac{dx_i}{dx_{p+1}} + Y_{p+2}\frac{dx_i}{dx_{p+2}} + \cdots Y_n \frac{dx_i}{dx_n} - Y_i. \qquad (13')$$
$$(i = 1, 2, \ldots p)$$

Let:

$$Y_i = \sum C_{i,\beta,\alpha_1,\alpha_2,\cdots \alpha_n} e^{\beta t \sqrt{-1}} x_1^{\alpha_1} x_2^{\alpha_2} \cdots x_n^{\alpha_n}$$

be any one of functions $Y_1, Y_2, \ldots Y_n$, as we had assumed above, and let us propose to calculate the p functions

$$x_1, x_2, \ldots x_p$$

in the form

$$x_i = \sum A_{i,\beta,\alpha_{p+1},\alpha_{p+2},\ldots \alpha_n} e^{\beta t \sqrt{-1}} x_{p+1}^{\alpha_{p+1}} x_{p+2}^{\alpha_{p+2}} \cdots x_n^{\alpha_n}. \qquad (15)$$

In order to calculate the coefficients A by iteration, we substitute the series (15) into Eq. (13') and align the left- and right-sides. We will have the following equation for calculating $A_{i,\beta,\alpha_{p+1},\alpha_{p+2},\ldots \alpha_n}$:

$$A_{i,\beta,\alpha_{p+1},\alpha_{p+2},\ldots \alpha_n}\left(\beta\sqrt{-1} + \alpha_{p+1}\lambda_{p+1} + \alpha_{p+2}\lambda_{p+2} + \cdots \alpha_n \lambda_n - \lambda_i\right)$$
$$= P[C, (-C'), A]$$

where $P[C,(-C'),A]$ is an integer polynomial with positive coefficients for the coefficients C from

$$Y_{p+1}, Y_{p+2}, \ldots Y_n,$$

for the coefficients C' from Y_i with the sign reversed and for the coefficients A already calculated.

So that none of the coefficients of A become infinite, we therefore need to first assume that there is a relationship of the following form between the λ:

$$\beta\sqrt{-1} + \alpha_{p+1}\lambda_{p+1} + \alpha_{p+2}\lambda_{p+2} + \cdots \alpha_n\lambda_n - \lambda_i = 0 \qquad (16)$$

where the α are positive integers and β is a positive or negative integer.

With that said, let λ' be a positive quantity which we will determine more completely in the following.

Next let:

$$Y'_i = \sum |C_{i,\beta,\alpha_1,\alpha_2,\ldots\alpha_n}| e^{\beta t\sqrt{-1}} x_1^{\alpha_1} x_2^{\alpha_2} \cdots x_n^{\alpha_n}$$

for

$$i = p+1, p+2, \ldots n$$

and

$$Y'_i = -\sum |C_{i,\beta,\alpha_1,\alpha_2,\ldots\alpha_n}| e^{\beta t\sqrt{-1}} x_1^{\alpha_1} x_2^{\alpha_2} \ldots x_n^{\alpha_n}$$

for

$$i = 1, 2, \ldots p.$$

We form the equations

$$\lambda' x_{p+1} \frac{dx_i}{dx_{p+1}} + \lambda' x_{p+2} \frac{dx_i}{dx_{p+2}} + \cdots \lambda' x_n \frac{dx_i}{dx_n} - \lambda' x_i$$
$$= Y'_{p+1} \frac{dx_i}{dx_{p+1}} + Y'_{p+2} \frac{dx_i}{dx_{p+2}} + \cdots Y'_n \frac{dx_i}{dx_n} - Y'_i. \qquad (13'')$$
$$(i = 1, 2, \ldots p)$$

We seek to satisfy Eq. (13″) using series of the following form:

$$x_i = \sum B_{i,\beta,\alpha_{p+1},\alpha_{p+2},\ldots\alpha_n} e^{\beta t\sqrt{-1}} x_{p+1}^{\alpha_{p+1}} x_{p+2}^{\alpha_{p+2}} \ldots x_n^{\alpha_n}. \qquad (15')$$

3 Application of the Calculation of Limits to Partial Differential Equations

The following equations will give us the coefficients B:

$$B_{i,\beta,\alpha_{p+1},\alpha_{p+2},\ldots\alpha_n}\left[\lambda'\left(\alpha_{p+1}+\alpha_{p+2}+\cdots\alpha_n-1\right)\right]=P[|C|,|C'|,B]$$

where $P[|C|,|C'|,B]$ differs from $P[C,(-C'),A]$ in that the coefficients C and C' in it are replaced by their magnitudes and the coefficients A by the corresponding B.

From this we conclude that all the B are positive and that each B is greater than the magnitude of the corresponding A.

A single condition is necessary for this; it is:

$$\lambda'\left(\alpha_{p+1}+\alpha_{p+2}+\cdots\alpha_n-1\right)<\left|\beta\sqrt{-1}+\alpha_{p+1}\lambda_{p+1}+\alpha_{p+2}\lambda_{p+2}+\cdots\alpha_n\lambda_n-\lambda_i\right|.$$

If this condition is satisfied, each of the terms from the series (15) will be smaller than the corresponding terms from the series (15′) and since this series converges, the series (15) will also converge.

For this it is sufficient that it be possible to find a positive quantity λ' which is small enough to always have:

$$\lambda'<\left|\frac{\beta\sqrt{-1}+\alpha_{p+1}\lambda_{p+1}+\cdots\alpha_n\lambda_n-\lambda_i}{\left(\alpha_{p+1}+\alpha_{p+2}+\cdots\alpha_n-1\right)}\right|.$$

This means, according to what we saw above, that neither of the two circumscribed *convex polygons*—the first with points $\lambda_{p+1},\lambda_{p+2},\ldots\lambda_n$ and $+\sqrt{-1}$, and the second with points $\lambda_{p+1},\lambda_{p+2},\ldots\lambda_n$ and $-\sqrt{-1}$—enclose the origin.

If therefore neither of the two convex polygons contains the origin, and if there is no relationship between the λ of form (16), then *Eq.* (13) will allow a specific integral of the following form:

$$\begin{aligned}x_1&=\varphi_1\left(x_{p+1},x_{p+2},\ldots x_n,t\right),\\ x_2&=\varphi_2\left(x_{p+1},x_{p+2},\ldots x_n,t\right),\\ &\vdots\\ x_p&=\varphi_p\left(x_{p+1},x_{p+2},\ldots x_n,t\right),\end{aligned}$$

where the φ are expandable in powers of $x_{p+1},x_{p+2},\ldots x_n$ and in sines and cosines of multiples of t.

With that said, consider the equations:

$$dt=\frac{dx_1}{X_1}=\frac{dx_2}{X_2}=\cdots\frac{dx_n}{X_n}. \tag{10″}$$

These equations have the same form as Eq. (10′); the only difference is that the λ do not have values which satisfy the sufficient conditions stated above for Eq. (13) to have a mapping integral.

We are going to propose that we find not the general solution for Eq. (10″), but a solution containing $n - p$ arbitrary constants.

Among Eq. (10″), I consider the following in particular:

$$\frac{dx_{p+1}}{dt} = X_{p+1}, \quad \frac{dx_{p+2}}{dt} = X_{p+2}, \quad \ldots \quad \frac{dx_n}{dt} = X_n. \tag{17}$$

Furthermore, I write the equations:

$$x_i = \varphi_i(x_{p+1}, x_{p+2}, \ldots x_n, t), \quad (i = 1, 2, \ldots p) \tag{18}$$

where the φ_i are mapping integrals of Eq. (13) defined above.

It is obvious that if $x_1, x_2, \ldots x_n$ are n functions of t which satisfy Eqs. (17) and (18), they will also satisfy Eq. (10″).

In Eq. (17) we substitute in place of $x_1, x_2, \ldots x_p$ their values (18) and these equations will become:

$$\frac{dx_{p+1}}{dt} = \lambda_{p+1} x_{p+1} + Z_{p+1}, \quad \frac{dx_{p+2}}{dt} = \lambda_{p+2} x_{p+2} + Z_{p+2}, \quad \ldots \quad \frac{dx_n}{dt} = \lambda_n x_{p+1} + Z_{p+1}, \tag{19}$$

where $Z_{p+1}, Z_{p+2}, \ldots Z_n$ are series expanded in powers of $x_{p+1}, x_{p+2}, \ldots x_n$ whose terms are at least second-order and whose coefficients are periodic functions of t.

This Eq. (19) has the same form as Eq. (10′); the general integral will therefore have the following form:

$$T'_{p+1} = K_{p+1} e^{\lambda_{p+1} t}, \quad T'_{p+2} = K_{p+2} e^{\lambda_{p+2} t}, \quad \ldots \quad T'_n = K_n e^{\lambda_n t},$$

where $K_{p+1}, K_{p+2}, \ldots K_n$ are constants of integration, and where $T'_{p+1}, T'_{p+2}, \ldots T'_n$ are series expanded in powers of x and the sines and cosines of multiples of t.

The equations:

$$\begin{aligned} T_i &= 0, \quad (i = 1, 2, \ldots p) \\ T'_q &= K_q e^{\lambda_q t}, \quad (q = p+1, p+2, \ldots n) \end{aligned} \tag{20}$$

therefore give us an integral of Eq. (10″) which depends on $n - p$ arbitrary constants $K_{p+1}, K_{p+2}, \ldots K_n$.

To get this integral in explicit form, these Eq. (20) must be solved for the $x_1, x_2, \ldots x_n$; in that way, we find:

3 Application of the Calculation of Limits to Partial Differential Equations

$$x_1 = \psi_1(t, K_{p+1}, \ldots K_n),$$
$$x_2 = \psi_2(t, K_{p+1}, \ldots K_n),$$
$$\vdots$$
$$x_n = \psi_n(t, K_{p+1}, \ldots K_n),$$

where the ψ are series expanded in powers of

$$K_{p+1}e^{\lambda_{p+1}t}, K_{p+2}e^{\lambda_{p+2}t}, \ldots K_n e^{\lambda_n t}$$

and in sines and cosines of multiples of t.

The series are convergent, provided that neither of the two circumscribed convex polygons—the first with points $\lambda_{p+1}, \lambda_{p+2}, \ldots \lambda_n$ and $+\sqrt{-1}$, and the second with points $\lambda_{p+1}, \lambda_{p+2}, \ldots \lambda_n$ and $-\sqrt{-1}$—contain the origin and that there is no relation between the λ of the form (16).

This proof brings out the analogy of this theorem with those that I stated in my thesis and in particular with this one:

In the neighborhood of a singular point, the solutions of a differential equation are expandable in powers of $t, t^{\lambda_1}, t^{\lambda_2}, \ldots t^{\lambda_n}$.

I had first demonstrated this theorem (which I later associated with the general ideas which inspired my thesis) by a rather different route in 45th *Cahier du Journal de l'École polytechnique* and Mr. Picard had been led to it independently by other considerations (*Comptes rendus*, 1878).

4 Integration of Linear Equations with Periodic Coefficients

It is known that a periodic function of x with period 2π is expandable in a series with the following form:

$$\begin{aligned}f(x) =\,& A_0 + A_1 \cos x + A_2 \cos 2x + \cdots A_n \cos nx + \cdots \\ & + B_1 \sin x + B_2 \sin 2x + \cdots B_n \sin nx + \cdots\end{aligned} \quad (1)$$

In the *Bulletin astronomique* (November 1886), I showed that if the function $f(x)$ is finite and continuous as well as its first $p-2$ derivatives and if its $p-1$ derivative is finite—but possibly discontinuous at a limited number of points—it is possible to find a positive number K such that the following hold however large n is:

$$|n^p A_n| < K, \quad |n^p B_n| < K.$$

If $f(x)$ is an analytic function, it will be finite and continuous and so will all its derivatives. It will therefore be possible to find a number K such that:

$$|n^2 A_n| < K, \quad |n^2 B_n| < K.$$

It follows from this that the series:

$$|A_0| + |A_1| + |A_2| + \cdots |A_n| + \cdots + |B_1| + |B_2| + \cdots |B_n| + \cdots$$

converges and consequently that the series (1) is absolutely and uniformly convergent.

With that established, consider a system of linear differential equations:

$$\begin{aligned}
\frac{dx_1}{dt} &= \varphi_{11} x_1 + \varphi_{12} x_2 + \cdots \varphi_{1n} x_n, \\
\frac{dx_2}{dt} &= \varphi_{21} x_1 + \varphi_{22} x_2 + \cdots \varphi_{2n} x_n, \\
&\vdots \\
\frac{dx_n}{dt} &= \varphi_{n1} x_1 + \varphi_{n2} x_2 + \cdots \varphi_{nn} x_n.
\end{aligned} \quad (2)$$

The n^2 coefficients φ_{ij} are periodic functions of t with period 2π.

Equation (2) therefore does not change when t changes to $t + 2\pi$. In light of that, let:

$$\begin{array}{cccc}
x_1 = \psi_{11}(t) & x_2 = \psi_{12}(t) & \cdots & x_n = \psi_{1n}(t), \\
x_1 = \psi_{21}(t) & x_2 = \psi_{22}(t) & \cdots & x_n = \psi_{2n}(t), \\
\vdots & \vdots & \ddots & \vdots \\
x_1 = \psi_{n1}(t) & x_2 = \psi_{n2}(t) & \cdots & x_n = \psi_{nn}(t)
\end{array} \quad (3)$$

be n linearly independent solutions of Eq. (2).

The equations do not change when t changes to $t + 2\pi$ and the n solutions will become:

$$\begin{array}{cccc}
x_1 = \psi_{11}(t+2\pi) & x_2 = \psi_{12}(t+2\pi) & \cdots & x_n = \psi_{1n}(t+2\pi), \\
x_1 = \psi_{21}(t+2\pi) & x_2 = \psi_{22}(t+2\pi) & \cdots & x_n = \psi_{2n}(t+2\pi), \\
\vdots & \vdots & \ddots & \vdots \\
x_1 = \psi_{n1}(t+2\pi) & x_2 = \psi_{n2}(t+2\pi) & \cdots & x_n = \psi_{nn}(t+2\pi).
\end{array}$$

There will therefore have to be linear combinations of the n solutions (3) such that we will have:

4 Integration of Linear Equations with Periodic Coefficients

$$\psi_{11}(t+2\pi) = A_{11}\psi_{11}(t) + A_{12}\psi_{21}(t) + \ldots A_{1n}\psi_{n1}(t),$$
$$\psi_{21}(t+2\pi) = A_{21}\psi_{11}(t) + A_{22}\psi_{21}(t) + \ldots A_{2n}\psi_{n1}(t),$$
$$\vdots$$
$$\psi_{n1}(t+2\pi) = A_{n1}\psi_{11}(t) + A_{n2}\psi_{21}(t) + \ldots A_{nn}\psi_{n1}(t),$$

where the A are constant coefficients.

Additionally, we will also have (with the same coefficients):

$$\psi_{12}(t+2\pi) = A_{11}\psi_{12}(t) + A_{12}\psi_{22}(t) + \cdots A_{1n}\psi_{n2}(t)$$

etc.

With that established, form the equation in S:

$$\begin{vmatrix} A_{11} - S & A_{12} & \cdots & A_{1n} \\ A_{21} & A_{22} - S & \cdots & A_{2n} \\ \vdots & \vdots & \ddots & \vdots \\ A_{n1} & A_{n2} & \cdots & A_{nn} - S \end{vmatrix} = 0. \qquad (5)$$

Let S_1 be one of the roots of this equation. According to the theory of linear substitutions, there will still be n constant coefficients

$$B_1, B_2, \ldots B_n$$

such that if one sets:

$$\theta_{11}(t) = B_1\psi_{11}(t) + B_2\psi_{21}(t) + \cdots B_n\psi_{n1}(t)$$

and likewise:

$$\theta_{1i}(t) = B_1\psi_{1i}(t) + B_2\psi_{2i}(t) + \cdots B_n\psi_{ni}(t)$$

one would have:

$$\theta_{11}(t+2\pi) = S_1\theta_{11}(t)$$

and likewise:

$$\theta_{1i}(t+2\pi) = S_1\theta_{1i}(t)$$

Set:

$$S_1 = e^{2\alpha_1 \pi}$$

which with the previous equation:

$$e^{-\alpha_1(t+2\pi)}\theta_{11}(t+2\pi) = S_1 e^{-2\alpha_1\pi}e^{-\alpha_1 t}\theta_{11}(t) = e^{-\alpha_1 t}\theta_{11}(t).$$

This equation expresses that:

$$e^{-\alpha_1 t}\theta_{11}(t)$$

is a periodic function that we will be able to expand in a trigonometric series:

$$\lambda_{11}(t)$$

If the periodic functions $\varphi_{ik}(t)$ are analytic, then the solutions to the differential Eq. (2) and $\lambda_{11}(t)$ will be also. The series $\lambda_{11}(t)$ will therefore be absolutely and uniformly convergent.

Similarly

$$e^{-\alpha_1 t}\theta_{1i}(t)$$

will be a periodic function that it will be possible to represent by trigonometric series:

$$\lambda_{1i}(t).$$

We therefore have a specific solution to Eq. (2) which is written:

$$x_1 = e^{\alpha_1 t}\lambda_{11}(t), \quad x_2 = e^{\alpha_1 t}\lambda_{12}(t), \quad \cdots \quad x_n = e^{\alpha_1 t}\lambda_{1n}(t) \qquad (6)$$

For each root of Eq. (5), there corresponds a solution of the form (6).

If all the roots of Eq. (5) are distinct, we will then have n linearly independent solutions of this form and the general solution will be written:

$$\begin{aligned} x_1 &= C_1 e^{\alpha_1 t}\lambda_{11}(t) + C_2 e^{\alpha_2 t}\lambda_{21}(t) + \cdots C_n e^{\alpha_n t}\lambda_{n1}(t), \\ x_2 &= C_1 e^{\alpha_1 t}\lambda_{12}(t) + C_2 e^{\alpha_2 t}\lambda_{22}(t) + \cdots C_n e^{\alpha_n t}\lambda_{n2}(t), \\ &\vdots \\ x_n &= C_1 e^{\alpha_1 t}\lambda_{1n}(t) + C_2 e^{\alpha_2 t}\lambda_{2n}(t) + \cdots C_n e^{\alpha_n t}\lambda_{nn}(t). \end{aligned} \qquad (7)$$

The C are constants of integration, the α are constants, and the λ are absolutely and uniformly convergent trigonometric series.

Now let us look at what happens when Eq. (5) has a double root, for example when $\alpha_1 = \alpha_2$. Go back to formula (7), in it make

$$C_3 = C_4 = \ldots C_n = 0$$

4 Integration of Linear Equations with Periodic Coefficients

and make α_2 approach α_1. It becomes:

$$x_1 = e^{\alpha_1 t}\left[C_1 \lambda_{11}(t) + C_2 e^{(\alpha_2-\alpha_1)t}\lambda_{21}(t)\right]$$

or by setting

$$C_1 = C_1' - C_2, \quad C_2 = \frac{C_2'}{\alpha_2 - \alpha_1},$$

it becomes:

$$x_1 = e^{\alpha_1 t}\left[C_1' \lambda_{11}(t) + C_2' \frac{e^{(\alpha_2-\alpha_1)t}\lambda_{21}(t) - \lambda_{11}(t)}{\alpha_2 - \alpha_1}\right].$$

It is clear that the difference

$$\lambda_{21}(t) - \lambda_{11}(t)$$

will become zero for $\alpha_2 = \alpha_1$. We can therefore set:

$$\lambda_{21}(t) = \lambda_{11}(t) + (\alpha_2 - \alpha_1)\lambda'(t).$$

It thus becomes:

$$x_1 = e^{\alpha_1 t}\left[C_1' \lambda_{11}(t) + C_2'\lambda_{11}\frac{e^{(\alpha_2-\alpha_1)t} - 1}{\alpha_2 - \alpha_1} + C_2'\lambda'(t)e^{(\alpha_2-\alpha_1)t}\right]$$

and at the limit (for $\alpha_2 = \alpha_1$);

$$x_1 = C_1' e^{\alpha_1 t}\lambda_{11}(t) + C_2' e^{\alpha_1 t}[t\lambda_{11} + \lim \lambda'(t)].$$

It will be seen that the limit of $\lambda'(t)$ for $\alpha_2 = \alpha_1$ is again an absolutely and uniformly convergent trigonometric series.

Thus the effect of the presence of a double root in Eq. (5) has been to introduce into the solution terms of the following form:

$$e^{\alpha_1 t}t\lambda(t),$$

where $\lambda(t)$ is a trigonometric series.

It can be seen without difficulty that a triple root would introduce terms of the form:

$$e^{\alpha_1 t}t^2\lambda(t)$$

and so on.

I am not going to stress all these details. These results are well known from the work of Floquet, Callandreau, Bruns, and Stieltjes and if I have provided the proof here in full for the general case it is because its extreme simplicity made it possible for me to do it in a few words.

Chapter 2
Theory of Integral Invariants

1 Various Properties of the Equations of Dynamics

Let F be a function of a double series of variables:

$$x_1, x_2, \ldots x_n, y_1, y_2, \ldots y_n$$

and of time t.

Suppose that we have differential equations:

$$\frac{dx_i}{dt} = \frac{dF}{dy_i}, \quad \frac{dy_i}{dt} = -\frac{dF}{dx_i}. \tag{1}$$

Consider two infinitesimally close solutions of these equations:

$$x_1, x_2, \ldots x_n, y_1, y_2, \ldots y_n,$$
$$x_1 + \xi_1, x_2 + \xi_2, \ldots x_n + \xi_n, y_1 + \eta_1, y_2 + \eta_2, \ldots y_n + \eta_n,$$

where the ξ and the η are small enough that their squares can be neglected.

The ξ and the η will then satisfy the linear differential equations:

$$\frac{d\xi_i}{dt} = \sum_k \frac{d^2 F}{dy_i dx_k} \xi_k + \sum_k \frac{d^2 F}{dy_i dy_k} \eta_k,$$
$$\frac{d\eta_i}{dt} = -\sum_k \frac{d^2 F}{dx_i dx_k} \xi_k - \sum_k \frac{d^2 F}{dx_i dy_k} \eta_k, \tag{2}$$

which are the perturbation equations of equations (1) (first-order Taylor series expansions).

Let ξ_i', η_i' be another solution of these linear equations such that:

$$\frac{d\xi_i'}{dt} = \sum_k \frac{d^2 F}{dy_i dx_k}\xi_k' + \sum_k \frac{d^2 F}{dy_i dy_k}\eta_k',$$
$$\frac{d\eta_i'}{dt} = -\sum_k \frac{d^2 F}{dx_i dx_k}\xi_k' - \sum_k \frac{d^2 F}{dx_i dy_k}\eta_k'. \qquad (2')$$

Multiply Eqs. (2) and (2'), respectively, by $\eta_i', -\xi_i', -\eta_i, \xi_i$ and add up all these equations, the result is:

$$\sum_i \left(\eta_i' \frac{d\xi_i}{dt} - \xi_i' \frac{d\eta_i}{dt} - \eta_i \frac{d\xi_i'}{dt} + \xi_i \frac{d\eta_i'}{dt} \right)$$
$$= \sum_i \sum_k \left(\xi_k \eta_i' \frac{d^2 F}{dy_i dx_k} + \eta_k \eta_i' \frac{d^2 F}{dy_i dy_k} + \xi_k \xi_i' \frac{d^2 F}{dx_i dx_k} + \eta_k \xi_i' \frac{d^2 F}{dx_i dy_k} \right)$$
$$- \sum_i \sum_k \left(\eta_i \xi_k' \frac{d^2 F}{dy_i dx_k} + \eta_i \eta_k' \frac{d^2 F}{dy_i dy_k} + \xi_i \xi_k' \frac{d^2 F}{dx_i dx_k} + \xi_i \eta_k' \frac{d^2 F}{dx_i dy_k} \right)$$

or

$$\sum_i \frac{d}{dt}[\eta_i' \xi_i - \xi_i' \eta_i] = 0$$

or finally

$$\eta_1' \xi_1 - \xi_1' \eta_1 + \eta_2' \xi_2 - \xi_2' \eta_2 + \ldots \eta_n' \xi_n - \xi_n' \eta_n = \text{const}. \qquad (3)$$

This is a relation which connects the two arbitrary solutions of the linear equations (2) to each other.

It is easy to find other analogous relations.

Consider for solutions of Eq. (2)

$$\xi_i, \xi_i', \xi_i'', \xi_i'''$$
$$\eta_i, \eta_i', \eta_i'', \eta_i'''.$$

Then consider the sum of the determinants:

$$\sum_i \sum_k \begin{vmatrix} \xi_i & \xi_i' & \xi_i'' & \xi_i''' \\ \eta_i & \eta_i' & \eta_i'' & \eta_i''' \\ \xi_k & \xi_k' & \xi_k'' & \xi_k''' \\ \eta_k & \eta_k' & \eta_k'' & \eta_k''' \end{vmatrix},$$

where the indices i and k vary from 1 to n. It can be verified without difficulty that this sum is again a constant.

1 Various Properties of the Equations of Dynamics

More generally, if the sum of determinants is formed using $2p$ solutions of Eq. (2):

$$\sum_{\alpha_1,\alpha_2,\ldots\alpha_p} \left|\xi_{\alpha_1}\eta_{\alpha_1}\xi_{\alpha_2}\eta_{\alpha_2}\ldots\xi_{\alpha_p}\eta_{\alpha_p}\right|, \quad (\alpha_1,\alpha_2,\ldots\alpha_p = 1,2,\ldots n)$$

this sum will be a constant.

In particular, the determinant formed by the values of the $2n$ quantities ξ and η in $2n$ solutions of Eq. (2) will be a constant.

Using these considerations it is possible to find a solution of Eq. (2) when an integral of them is known and vice versa.

Suppose in fact that

$$\xi_i = \alpha_i, \quad \eta_i = \beta_i$$

is a specific solution of Eq. (2) and designate an arbitrary solution of the same equations by ξ_i and η_i. We will then have:

$$\sum \xi_i \beta_i - \eta_i \alpha_i = \text{const.}$$

which will be an integral of Eq. (2).

And the other-way-around, let

$$\sum A_i \xi_i + \sum B_i \eta_i = \text{const.}$$

be an integral of Eq. (2), we will then have:

$$\sum_i \frac{dA_i}{dt}\xi_i + \sum_i \frac{dB_i}{dt}\eta_i + \sum_i A_i \left[\sum_k \frac{d^2 F}{dy_i dx_k}\xi_k + \sum_k \frac{d^2 F}{dy_i dy_k}\eta_k\right]$$

$$- \sum_i B_i \left[\sum_k \frac{d^2 F}{dx_i dx_k}\xi_k + \sum_k \frac{d^2 F}{dx_i dy_k}\eta_k\right] = 0,$$

hence by aligning terms:

$$\frac{dA_i}{dt} = \sum_k \frac{d^2 F}{dy_i dx_k}A_k + \sum_k \frac{d^2 F}{dy_i dy_k}B_k,$$

$$\frac{dB_i}{dt} = -\sum_k \frac{d^2 F}{dx_i dx_k}A_k - \sum_k \frac{d^2 F}{dy_i dy_k}B_k,$$

which shows that:

$$\xi_i = B_i, \quad \eta_i = -A_i$$

is a specific solution of Eq. (2).

If now:

$$\Phi(x_i, y_i, t) = \text{const.}$$

is an integral of Eq. (1), then

$$\sum \frac{d\Phi}{dx_i} \xi_i + \sum \frac{d\Phi}{dy_i} \eta_i = \text{const.}$$

will be an integral of Eq. (2) and consequently:

$$\xi_i = \frac{d\Phi}{dy_i}, \quad \eta_i = -\frac{d\Phi}{dx_i}$$

will be a specific solution of these equations.

If $\Phi = \text{const.}$ and $\Phi_1 = \text{const.}$ are two integrals of Eq. (1), then we will have

$$\sum \left(\frac{d\Phi}{dx_i} \frac{d\Phi_1}{dy_i} - \frac{d\Phi}{dy_i} \frac{d\Phi_1}{dx_i} \right) = \text{const.}$$

This is Poisson's theorem.

Consider the specific case where the x designate rectangular coordinates of n spatial points; we will designate them using double index notation:

$$x_{1i}, \quad x_{2i}, \quad x_{3i},$$

where the first index refers to the three rectangular coordinates and the second index to the n material points. Let m_i be the mass of material point i. We will then have:

$$m_i \frac{d^2 x_{ki}}{dt^2} = \frac{dV}{dx_{ki}},$$

where V is the potential energy.

We will then have the equation for the conservation of energy:

$$F = \sum \frac{m_i}{2} \left(\frac{dx_{ki}}{dt} \right)^2 - V = \text{const.}$$

Next set:

$$y_{ki} = m_i \frac{dx_{ki}}{dt}$$

1 Various Properties of the Equations of Dynamics

hence:

$$F = \sum \frac{y_{ki}^2}{2m_i} - V = \text{const.} \tag{4}$$

and

$$\frac{dx_{ki}}{dt} = \frac{dF}{dy_{ki}}, \quad \frac{dy_{ki}}{dt} = -\frac{dF}{dx_{ki}}. \tag{1'}$$

Let:

$$x_{ki} = \varphi_{ki}(t), \quad y_{ki} = m_i \varphi'_{ki}(t) \tag{5}$$

be a solution of this Eq. (1') and another solution be:

$$x_{ki} = \varphi_{ki}(t+h), \quad y_{ki} = m_i \varphi'_{ki}(t+h),$$

where h is an arbitrary constant.

By thinking of h as infinitesimal, a solution of Eq. (2') can be obtained which correspond to (1') as Eq. (2) correspond to (1):

$$\xi_{ki} = h\varphi'_{ki}(t) = h\frac{y_{ki}}{m_i}, \quad \eta_{ki} = hm_i\varphi''_{ki}(t) = h\frac{dV}{dx_{ki}},$$

where h designates a very small constant factor which can be dropped when only linear Eq. (2') are considered.

Knowing a solution:

$$\xi = \frac{y}{m}, \quad \eta = \frac{dV}{dx}$$

of these equations, an integral can be deduced:

$$\sum \frac{y\eta}{m} - \sum \frac{dV}{dx}\xi = \text{const.}$$

But this same integral can be obtained very easily by differentiating the energy conservation Eq. (4).

If the material points are free of any outside action, another solution can be deduced from solution (5):

$$\begin{aligned}
x_{1i} &= \varphi_{1i}(t) + h + kt, & y_{1i} &= m_i\varphi'_{1i}(t) + m_i k,\\
x_{2i} &= \varphi_{2i}(t), & y_{2i} &= m_i\varphi'_{2i}(t),\\
x_{3i} &= \varphi_{3i}(t), & y_{3i} &= m_i\varphi'_{3i}(t),
\end{aligned}$$

where h and k are arbitrary constants. By thinking of these constants as infinitesimally small, we get two solutions of Eq. (2'):

$$\xi_{1i} = 1, \xi_{2i} = \xi_{3i} = \eta_{1i} = \eta_{2i} = \eta_{3i} = 0,$$
$$\xi_{1i} = t, \xi_{2i} = \xi_{3i} = \eta_{2i} = \eta_{3i} = 0, \eta_{1i} = m_i.$$

Thus two integrals of (2') can be obtained:

$$\sum_i \eta_{1i} = \text{const.},$$

$$\sum_i \eta_{1i} t - \sum_i m_i \xi_{1i} = \text{const.}$$

These integrals can also be obtained by differentiating the equations of motion of the center of gravity:

$$\sum_i m_i x_{1i} = t \sum_i y_{1i} + \text{const.},$$

$$\sum_i y_{1i} = \text{const.}$$

By rotating the solution (5) through an angle ω around the z-axis, another solution is obtained:

$$x_{1i} = \varphi_{1i} \cos\omega - \varphi_{2i} \sin\omega, \quad \frac{y_{1i}}{m_i} = \varphi'_{1i} \cos\omega - \varphi'_{2i} \sin\omega,$$
$$x_{2i} = \varphi_{1i} \sin\omega + \varphi_{2i} \cos\omega, \quad \frac{y_{2i}}{m_i} = \varphi'_{1i} \sin\omega + \varphi'_{2i} \cos\omega,$$
$$x_{3i} = \varphi_{3i}, \quad \frac{y_{3i}}{m_i} = \varphi'_{3i}.$$

By regarding ω as infinitesimally small, we find a solution of (2')

$$\xi_{1i} = -x_{2i}, \quad \eta_{1i} = -y_{2i},$$
$$\xi_{2i} = x_{1i}, \quad \eta_{2i} = y_{1i},$$
$$\xi_{3i} = 0, \quad \eta_{3i} = 0,$$

and hence the integral for (2')

$$\sum_i (x_{1i}\eta_{2i} - y_{1i}\xi_{2i} - x_{2i}\eta_{1i} + y_{2i}\xi_{1i}) = \text{const.}$$

1 Various Properties of the Equations of Dynamics

that can also be obtained by differentiating the integral of the areas from (1')

$$\sum_i (x_{1i} y_{2i} - x_{2i} y_{1i}) = \text{const.}$$

Now suppose that the function V is homogeneous and of degree -1 in x which is the case in nature.

Equation (1') does not change when t is multiplied by λ^3, the x by λ^2, and the y by λ^{-1}, where λ is an arbitrary constant. From the solution (4), the following solution can be deduced:

$$x_{ki} = \lambda^2 \varphi_{ki}\left(\frac{t}{\lambda^3}\right) \quad y_{ki} = \lambda^{-1} m_i \varphi'_{ki}\left(\frac{t}{\lambda^3}\right).$$

If λ is thought of as very close to unity, we will get the following results for the solutions of Eq. (2')

$$\xi_{ki} = 2\varphi_{ki} - 3t\varphi'_{ki}, \quad \eta_{ki} = -m_i \varphi'_{ki} - 3m_i t \varphi''_{ki},$$

or

$$\xi_{ki} = 2x_{ki} - 3t\frac{y_{ki}}{m_i}, \quad \eta_{ki} = -y_{ki} - 3t\frac{\mathrm{d}V}{\mathrm{d}x_{ki}}, \tag{6}$$

and hence the following integral for Eq. (2'), which, unlike those which we have considered up to here, cannot be obtained by differentiating a known integral of Eq. (1'):

$$\sum (2x_{ki}\eta_{ki} + y_{ki}\xi_{ki}) = 3t\left[\sum\left(\frac{y_{ki}\eta_{ki}}{m_i} - \frac{\mathrm{d}V}{\mathrm{d}x_{ki}}\xi_{ki}\right)\right] + \text{const.}$$

2 Definitions of Integral Invariants

Consider a system of differential equations:

$$\frac{\mathrm{d}x_i}{\mathrm{d}t} = X_i,$$

where X_i is a given function of $x_1, x_2, \ldots x_n$. If we have:

$$F(x_1, x_2, \ldots x_n) = \text{const.},$$

then this relationship is called an integral of the given equations. The left-hand side of this relationship can be called an invariant because it is not altered when the x_i are increased by infinitesimal increases $\mathrm{d}x_i$ compatible with the differential equations.

Now let

$$x'_1, x'_2, \ldots x'_n$$

be another solution of the same differential equations, such that we have:

$$\frac{dx'_i}{dt} = X'_i$$

where X'_i is a function formed with $x'_1, x'_2, \ldots x'_n$ as X_i was formed with $x_1, x_2, \ldots x_n$.

It is possible that there could be a relationship of the following form between the $2n$ quantities x and x':

$$F_1(x_1, x_2, \ldots x_n, x'_1, x'_2, \ldots x'_n) = \text{const.}$$

The left-hand side, F_1, could also be called an invariant of our differential equations, because instead of depending on a single solution of these equations, it will depend on two solutions.

It can be assumed that $x_1, x_2, \ldots x_n$ represents the coordinates of a point in n dimensional space and that the given differential equations define the laws of motion of this point. If we think about the two solutions of these equations, there are two different moving points, moving under a single law defined by our differential equations. The invariant F_1 will then be a function of the coordinates of these two points and the invariant will retain its initial value during the motion of these two points.

Similarly, instead of two moving points, three or even a large number of moving points could obviously be considered.

Now assume that infinitely many moving points are being considered and that the initial positions of these points form a specific arc of curve C in the n dimensional space.

When we are given the initial position of a moving point and the differential equations which define its laws of motion, the position of the point at an arbitrary moment is then completely determined.

If we therefore know that our moving points, infinitely many, form an arc C at the origin of time, we will know their positions at an arbitrary time t and we will see that the moving points at the moment t form a new arc C' in the n dimensional space. We therefore have an arc of curve which moves while changing shape because its various points move according to the laws defined by the given differential equations.

Now assume that during this motion and this deformation, the following integral:

$$\int (Y_1 dx_1 + Y_2 dx_2 + \ldots Y_n dx_n) = \int \sum Y_i dx_i$$

2 Definitions of Integral Invariants

(where the Y are given functions of the x and which extends the entire length of the curve) does not change value. This integral will again be an invariant for our differential equations, no longer depending on one, two or three points, but on infinitely many moving points. To indicate what its shape is, I will call it an integral invariant.

Similarly it can be imagined that an integral of the following form:

$$\int \sqrt{\sum Y_{ik} dx_i dx_k},$$

over the entire arc of the curve could remain invariant; this again would be an integral invariant.

Integral invariants can also be imagined which are defined by double or multiple integrals.

Imagine that we are considering a fluid in continuous motion such that the three components X, Y, Z of the speed of an arbitrary molecule are given functions of the three coordinates x, y, z of this molecule. Then it would be possible to state that the laws of motion of an arbitrary fluid molecule are defined by the differential equations:

$$\frac{dx}{dt} = X, \quad \frac{dy}{dt} = Y, \quad \frac{dz}{dt} = Z.$$

It is known that the partial differential equation

$$\frac{dX}{dx} + \frac{dY}{dy} + \frac{dZ}{dz} = 0$$

expresses that the fluid is incompressible. Therefore assume that the functions X, Y, Z satisfy this equation and consider an ensemble of molecules occupying a specific volume at the origin of time. The molecules will move, but because the fluid is incompressible the volume that they occupy will remain unchanged. In other words the volume, meaning the triple integral:

$$\iiint dx dy dz$$

will be an integral invariant. More generally, if we consider the equations:

$$\frac{dx_i}{dt} = X_i \quad (i = 1, 2, \ldots n)$$

and we have the relationship:

$$\sum_{i=1}^{n} \frac{dX_i}{dx_i} = 0,$$

the nth order integral

$$\int dx_1 dx_2 \ldots dx_n$$

which I will continue to call the volume, will be an integral invariant.

This is what will happen in particular for the general equations of dynamics; because on consideration of these equations:

$$\frac{dx_i}{dt} = \frac{dF}{dy_i}, \frac{dy_i}{dt} = -\frac{dF}{dx_i},$$

it is easy to see that

$$\sum \frac{d\left(\frac{dF}{dy_i}\right)}{dx_i} + \sum \frac{d\left(-\frac{dF}{dx_i}\right)}{dy_i} = 0$$

But as it relates to the general equations of dynamics, there is in addition to the volume, another integral invariant that will be even more useful to us. We have in fact seen that:

$$\sum (\xi_i \eta_i' - \xi_i' \eta_i) = \text{const.}$$

Which translated into our new language means that the double integral

$$\iint \sum_i dx_i dy_i$$

is an integral invariant, as I will prove below.

To express this result in another way, take the case of the n-body problem.

We will represent the state of the system of n bodies by the position of $3n$ points in a plane. The abscissa of the first point will be the x of the first body and the ordinate the projection on the x-axis of the momentum of this body; the abscissa of the second point will be the y of the same body and the ordinate the projection on the y-axis of its momentum and so on.

Imagine a double infinity of initial states of the system. A position of our $3n$ points corresponds to each of them and if all of these states are considered, it will be seen that the $3n$ points fill $3n$ plane areas.

If the system now moves according to the law of gravitational attraction the $3n$ points which represent its state are also going to move; the plane areas that I just defined are going to deform, but *their sum will remain constant*.

The theorem on the conservation of volume is just one consequence of the preceding.

In the case of the n-body problem there is another integral invariant to which I want to draw attention.

2 Definitions of Integral Invariants

Consider a single infinity of initial positions of the system which forms an arc of curve in the $6n$ dimensional space. Let C_0 and C_1 be the values of the constant of total energy at two ends of this arc. I will demonstrate later that the expression

$$\int \sum (2x_i dy_i + y_i dx_i) + 3(C_1 - C_0)t$$

(where the integral is along the arc of the entire curve and where the time does not enter if $C_0 = C_1$) is again an integral invariant; it is furthermore possible to easily deduce the other integral invariants which were covered above.

We will state that an integral invariant is of first-order, second-order, ..., or of nth order according to whether it is a single, double, ..., or n times integral.

Among the integral invariants we will distinguish the *positive invariants* that we will define as follows.

The nth order integral invariant:

$$\int M dx_1 dx_2 \ldots dx_n$$

will be a positive invariant in some domain, if M is a function of $x_1, x_2, \ldots x_n$ which remains positive, finite and one-to-one in this domain.

I still need to prove the various results which I just stated; this proof can be done by a very simple calculation.

Let:

$$\frac{dx_1}{dt} = X_1, \quad \frac{dx_2}{dt} = X_2, \quad \ldots \quad \frac{dx_n}{dt} = X_n \tag{1}$$

be a system of differential equations where $X_1, X_2, \ldots X_n$ are functions of $x_1, x_2, \ldots x_n$ such that:

$$\frac{dX_1}{dx_1} + \frac{dX_2}{dx_2} + \cdots \frac{dX_n}{dx_n} = 0. \tag{2}$$

Let there be a solution to this system of equations which depends on n arbitrary constants:

$$\alpha_1, \alpha_2, \ldots \alpha_n.$$

This solution will be written:

$$x_1 = \varphi_1(t, \alpha_1, \alpha_2, \ldots \alpha_n),$$
$$x_2 = \varphi_2(t, \alpha_1, \alpha_2, \ldots \alpha_n),$$
$$\vdots$$
$$x_n = \varphi_n(t, \alpha_1, \alpha_2, \ldots \alpha_n).$$

It is a matter of demonstrating that the integral

$$J = \int dx_1 dx_2 \ldots dx_n = \int \Delta d\alpha_1 d\alpha_2 \ldots d\alpha_n$$

where

$$\Delta = \begin{vmatrix} \dfrac{dx_1}{d\alpha_1} & \dfrac{dx_2}{d\alpha_1} & \cdots & \dfrac{dx_n}{d\alpha_1} \\ \dfrac{dx_1}{d\alpha_2} & \dfrac{dx_2}{d\alpha_2} & \cdots & \dfrac{dx_n}{d\alpha_2} \\ \vdots & \vdots & \ddots & \vdots \\ \dfrac{dx_1}{d\alpha_n} & \dfrac{dx_2}{d\alpha_n} & \cdots & \dfrac{dx_n}{d\alpha_n} \end{vmatrix}$$

is a constant.

In fact we have:

$$\frac{dJ}{dt} = \int \frac{d\Delta}{dt} d\alpha_1 d\alpha_2 \ldots d\alpha_n$$

and

$$\frac{d\Delta}{dt} = \Delta_1 + \Delta_2 + \ldots \Delta_n,$$

where Δ_k is the determinant Δ in which the kth column

$$\begin{array}{c} \dfrac{dx_k}{d\alpha_1} \\ \dfrac{dx_k}{d\alpha_2} \\ \vdots \\ \dfrac{dx_k}{d\alpha_n} \end{array} \quad \text{is replaced by} \quad \begin{array}{c} \dfrac{d^2 x_k}{d\alpha_1 dt} \\ \dfrac{d^2 x_k}{d\alpha_2 dt} \\ \vdots \\ \dfrac{d^2 x_k}{d\alpha_n dt} \end{array}.$$

But we have

$$\frac{dx_k}{dt} = X_k,$$

hence

$$\frac{d^2 x_k}{d\alpha_i dt} = \frac{dX_k}{dx_1} \frac{dx_1}{d\alpha_i} + \frac{dX_k}{dx_2} \frac{dx_2}{d\alpha_i} + \ldots \frac{dX_k}{dx_n} \frac{dx_n}{d\alpha_i}.$$

2 Definitions of Integral Invariants

We deduce from that:

$$\Delta_k = \Delta \frac{dX_k}{dx_k},$$

hence

$$\frac{dJ}{dt} = \int (\Delta_1 + \Delta_2 + \cdots \Delta_n) d\alpha_1 d\alpha_2 \ldots d\alpha_n$$
$$= \int \left(\frac{dX_1}{dx_1} + \frac{dX_2}{dx_2} + \cdots \frac{dX_n}{dx_n} \right) \Delta d\alpha_1 d\alpha_2 \ldots d\alpha_n = 0.$$

Which was to be proved.

Now suppose that instead of the relation (2) we had:

$$\frac{dMX_1}{dx_1} + \frac{dMX_2}{dx_2} + \cdots \frac{dMX_n}{dx_n} = 0 \qquad (2')$$

where M is an arbitrary function of $x_1, x_2, \ldots x_n$.

I state that:

$$J = \int M dx_1 dx_2 \ldots dx_n = \int M \Delta d\alpha_1 d\alpha_2 \ldots d\alpha_n$$

is a constant.

In fact we have:

$$\frac{dJ}{dt} = \int \left(\Delta \frac{dM}{dt} + M \frac{d\Delta}{dt} \right) d\alpha_1 d\alpha_2 \ldots d\alpha_n.$$

It must be shown that:

$$\Delta \frac{dM}{dt} + M \frac{d\Delta}{dt} = 0.$$

In fact we have [because of Eq. (1)]

$$\frac{dM}{dt} = X_1 \frac{dM}{dx_1} + X_2 \frac{dM}{dx_2} + \cdots X_n \frac{dM}{dx_n}$$

and (according to what we just saw):

$$\frac{d\Delta}{dt} = \Delta \left(\frac{dX_1}{dx_1} + \frac{dX_2}{dx_2} + \cdots \frac{dX_n}{dx_n} \right).$$

It therefore follows that:

$$\Delta \frac{dM}{dt} + M \frac{d\Delta}{dt} = \Delta \left(\frac{dMX_1}{dx_1} + \frac{dMX_2}{dx_2} + \cdots \frac{dMX_n}{dx_n} \right) = 0.$$

Which was to be proved.

Now move on to the equations of dynamics.

Let the equations be:

$$\frac{dx_i}{dt} = \frac{dF}{dy_i}, \quad \frac{dy_i}{dt} = -\frac{dF}{dx_i}. \quad (i = 1, 2, \ldots n) \quad (1')$$

Let there be a solution containing two arbitrary constants α and β and written:

$$x_i = \varphi_i(t, \alpha, \beta)$$
$$y_i = \psi_i(t, \alpha, \beta).$$

I state that:

$$J = \int (dx_1 dy_1 + dx_2 dy_2 + \cdots dx_n dy_n) = \int \sum_{i=1}^{n} \left(\frac{dx_i}{d\alpha} \frac{dy_i}{d\beta} - \frac{dx_i}{d\beta} \frac{dy_i}{d\alpha} \right) d\alpha d\beta$$

is a constant.

It follows in fact that:

$$\frac{dJ}{dt} = \int \sum \left(\frac{d^2 x_i}{dt d\alpha} \frac{dy_i}{d\beta} + \frac{d^2 y_i}{dt d\beta} \frac{dx_i}{d\alpha} - \frac{d^2 x_i}{dt d\beta} \frac{dy_i}{d\alpha} - \frac{d^2 y_i}{dt d\alpha} \frac{dx_i}{d\beta} \right) d\alpha d\beta.$$

It then follows:

$$\frac{d^2 x_i}{dt d\alpha} = \sum_k \frac{d^2 F}{dy_i dx_k} \frac{dx_k}{d\alpha} + \sum_k \frac{d^2 F}{dy_i dy_k} \frac{dy_k}{d\alpha},$$

$$\frac{d^2 x_i}{dt d\beta} = \sum_k \frac{d^2 F}{dy_i dx_k} \frac{dx_k}{d\beta} + \sum_k \frac{d^2 F}{dy_i dy_k} \frac{dy_k}{d\beta},$$

$$\frac{d^2 y_i}{dt d\alpha} = -\sum_k \frac{d^2 F}{dx_i dx_k} \frac{dx_k}{d\alpha} - \sum_k \frac{d^2 F}{dx_i dy_k} \frac{dy_k}{d\alpha},$$

$$\frac{d^2 y_i}{dt d\beta} = -\sum_k \frac{d^2 F}{dx_i dx_k} \frac{dx_k}{d\beta} - \sum_k \frac{d^2 F}{dx_i dy_k} \frac{dy_k}{d\beta}.$$

2 Definitions of Integral Invariants

From that we conclude that:

$$\sum \left(\frac{d^2 x_i}{dt d\alpha} \frac{dy_i}{d\beta} - \frac{d^2 y_i}{dt d\alpha} \frac{dx_i}{d\beta} \right)$$
$$= \sum \sum \left(\frac{d^2 F}{dy_i dx_k} \frac{dx_k}{d\alpha} \frac{dy_i}{d\beta} + \frac{d^2 F}{dy_i dy_k} \frac{dy_k}{d\alpha} \frac{dy_i}{d\beta} + \frac{d^2 F}{dx_i dx_k} \frac{dx_k}{d\alpha} \frac{dx_i}{d\beta} + \frac{d^2 F}{dx_i dy_k} \frac{dx_i}{d\beta} \frac{dy_k}{d\alpha} \right).$$

The right-hand side of the equation does not change on permuting α and β, and therefore we have:

$$\sum \left(\frac{d^2 x_i}{dt d\alpha} \frac{dy_i}{d\beta} - \frac{d^2 y_i}{dt d\alpha} \frac{dx_i}{d\beta} \right) = \sum \left(\frac{d^2 x_i}{dt d\beta} \frac{dy_i}{d\alpha} - \frac{d^2 y_i}{dt d\beta} \frac{dx_i}{d\alpha} \right).$$

This equality expresses that the quantity under the integral sign in the expression for dJ/dt is zero and consequently that

$$\frac{dJ}{dt} = 0.$$

Which was to be proved.

It remains to consider the last of the integral invariants which comes up in the case of the n-body problem.

Return to the equations of dynamics, but by setting:

$$F = T + U,$$

where T depends only on y and U only on x. Additionally, T is homogeneous and second-degree and U homogeneous and -1 degree.

Take a solution

$$x_i = \varphi_i(t, \alpha), \quad y_i = \psi_i(t, \alpha)$$

which depends solely on a single arbitrary constant, α.

Consider the single integral:

$$J = \int \sum \left(2x_i \frac{dy_i}{d\alpha} + y_i \frac{dx_i}{d\alpha} \right) d\alpha + 3(C_1 - C_0) t,$$

where C_1 and C_0 are constant values of the function F at the ends of the arc along which the integral is calculated.

It follows that:

$$\frac{dJ}{dt} = \int \sum \left(2\frac{dx_i}{dt}\frac{dy_i}{d\alpha} + \frac{dy_i}{dt}\frac{dx_i}{d\alpha} + 2x_i\frac{d^2 y_i}{dt d\alpha} + y_i\frac{d^2 x_i}{dt d\alpha} \right) d\alpha + 3(C_1 - C_0).$$

It follows that:

$$\frac{dx_i}{dt} = \frac{dF}{dy_i} = \frac{dT}{dy_i}, \quad \frac{dy_i}{dt} = -\frac{dU}{dx_i},$$

$$\frac{d^2 x_i}{dt d\alpha} = \sum_k \frac{d^2 T}{dy_i dy_k}\frac{dy_k}{d\alpha}, \quad \frac{d^2 y_i}{dt d\alpha} = -\sum_k \frac{d^2 U}{dx_i dx_k}\frac{dx_k}{d\alpha},$$

hence

$$\frac{dJ}{dt} = \int \sum \sum \left(2\frac{dT}{dy_i}\frac{dy_i}{d\alpha} + y_i\frac{d^2 T}{dy_i dy_k}\frac{dy_k}{d\alpha} - \frac{dU}{dx_i}\frac{dx_i}{d\alpha} - 2x_i\frac{d^2 U}{dx_i dx_k}\frac{dx_k}{d\alpha} \right) d\alpha + 3(C_1 - C_0).$$

But because of the homogeneous function theorem we have:

$$\sum_i y_i \frac{d^2 T}{dy_i dy_k} = \frac{dT}{dy_k}, \quad \sum_i x_i \frac{d^2 U}{dx_i dx_k} = -2\frac{dU}{dx_k},$$

hence

$$\frac{dJ}{dt} = \int \sum \left(3\frac{dT}{dy_i}\frac{dy_i}{d\alpha} + 3\frac{dU}{dx_i}\frac{dx_i}{d\alpha} \right) d\alpha + 3(C_1 - C_0)$$

or

$$\frac{dJ}{dt} = 3\int (dT + dU) + 3(C_1 - C_0).$$

However, according to the definition of C_1 and C_0 we have

$$C_0 - C_1 = \int dF = \int (dT + dU).$$

It therefore follows that:

$$\frac{dJ}{dt} = 0.$$

Which was to be proved.

3 Transformation of Integral Invariants

Return to our differential equations:

$$\frac{dx_1}{dt} = X_1, \quad \frac{dx_2}{dt} = X_2, \quad \ldots \quad \frac{dx_n}{dt} = X_n \tag{1}$$

and assume that we have:

$$\frac{d(MX_1)}{dx_1} + \frac{d(MX_2)}{dx_2} + \cdots \frac{d(MX_n)}{dx_n} = 0, \tag{2}$$

such that the nth order integral

$$J = \int M dx_1 dx_2 \ldots dx_n$$

is an integral invariant.

Change variables by setting:

$$\begin{aligned} x_1 &= \psi_1(z_1, z_2, \ldots z_n), \\ x_2 &= \psi_2(z_1, z_2, \ldots z_n), \\ &\vdots \\ x_n &= \psi_n(z_1, z_2, \ldots z_n), \end{aligned} \tag{3}$$

and call Δ the Jacobian determinant of the n functions ψ relative to the n variables z.

After the change of variables we will have:

$$J = \int M\Delta dz_1 dz_2 \ldots dz_n.$$

If the invariant was positive before the change of variables, it will remain positive after this change, provided that Δ is always positive, finite, and one-to-one.

Since by permuting two of the variables z, the sign of Δ changes; it will be sufficient for us to assume that Δ always has the same sign or that it is never zero. It will additionally always need to be finite and one-to-one. This will happen if the change of variables (3) is bijective; meaning if, in the domain in consideration, the x are one-to-one functions of z and the z one-to-one functions of x.

Thus after a bijective change of variables, the positive invariants remain positive.

Here is an interesting specific case:

Suppose that an integral of Eq. (1) is known:

$$F(x_1, x_2, \ldots x_n) = C.$$

Take for new variables both $z_n = C$ and also $n - 1$ other variables $z_1, z_2, \ldots z_{n-1}$. It will often happen that $z_1, z_2, \ldots z_{n-1}$ can be chosen such that this change of variables is bijective in the domain in consideration.

After the change of variables, Eq. (1) becomes:

$$\frac{dz_1}{dt} = Z_1, \quad \frac{dz_2}{dt} = Z_2, \quad \ldots \quad \frac{dz_{n-1}}{dt} = Z_{n-1}, \quad \frac{dz_n}{dt} = Z_n = 0, \qquad (4)$$

where $Z_1, Z_2, \ldots Z_{n-1}$ are known functions of $z_1, z_2, \ldots z_n$. If the constant $C = z_n$ is regarded as a given of the problem, the equations are reduced to order $n-1$ and are written:

$$\frac{dz_1}{dt} = Z_1, \quad \frac{dz_2}{dt} = Z_2, \quad \ldots \quad \frac{dz_{n-1}}{dt} = Z_{n-1}, \qquad (4')$$

the functions Z now depend only on $z_1, z_2, \ldots z_{n-1}$ because z_n was replaced there by its numeric value.

If there is a positive invariant of Eq. (1)

$$J = \int M dx_1 dx_2 \ldots dx_n,$$

then Eq. (4) will also have a positive invariant:

$$J = \int \mu dz_1 dz_2 \ldots dz_{n-1} dz_n.$$

I now state that Eq. (4') which is of order $n-1$ also have a positive integral invariant which must be of order $n-1$.

In fact, stating that J is an integral invariant amounts to stating that

$$\frac{d(\mu Z_1)}{dz_1} + \frac{d(\mu Z_2)}{dz_2} + \ldots \frac{d(\mu Z_n)}{dz_n} = 0$$

or because Z_n is zero,

$$\frac{d(\mu Z_1)}{dz_1} + \frac{d(\mu Z_2)}{dz_2} + \ldots \frac{d(\mu Z_{n-1})}{dz_{n-1}} = 0,$$

which proves that the $n-1$ order integral

$$\int \mu dz_1 dz_2 \ldots dz_{n-1}$$

is an invariant for Eq. (4').

Up till now we have applied the changes of variables to the unknown functions $x_1, x_2, \ldots x_n$, but we have kept time t which is our independent variable. We are now going to assume that we set:

3 Transformation of Integral Invariants

$$t = \varphi(t_1)$$

and that we take t_1 as the new independent variable.
Equation (1) then become:

$$\frac{dx_i}{dt_1} = X'_i = X_i \frac{d\varphi}{dt_1} = X_i \frac{dt}{dt_1} \quad (i = 1, 2, \ldots n) \tag{5}$$

If Eq. (1) has an nth order integral invariant

$$J = \int M dx_1 dx_2 \ldots dx_n$$

then it will be true that

$$\sum \frac{d}{dx_i}(MX_i) = 0,$$

which can be written

$$\sum \frac{d}{dx_i}\left(M \frac{dt_1}{dt} X'_i\right) = 0.$$

Which shows that

$$\int M \frac{dt_1}{dt} dx_1 dx_2 \ldots dx_n$$

is an integral invariant of Eq. (5).

For this transformation to be useful, it is necessary that t and t_1 be related such that dt_1/dt can be regarded as a known, finite, continuous, and one-to-one function of $x_1, x_2, \ldots x_n$.

Suppose for example that we take for new independent variable:

$$x_n = t_1.$$

It then follows that

$$\frac{dt_1}{dt} = X_n$$

and Eq. (5) are written

$$\frac{dx_1}{dt_1} = \frac{X_1}{X_n}, \quad \frac{dx_2}{dt_1} = \frac{X_2}{X_n}, \quad \ldots \quad \frac{dx_{n-1}}{dt_1} = \frac{X_{n-1}}{X_n}, \quad \frac{dx_n}{dt_1} = 1,$$

and they allow as integral invariant:

$$\int MX_n dx_1 dx_2 \ldots dx_n.$$

Similarly, if we take for new independent variable:

$$t_1 = \Theta(x_1, x_2, \ldots x_n),$$

where Θ is an arbitrary function of $x_1, x_2, \ldots x_n$, the new integral invariant will be written:

$$\int M\left(\frac{d\Theta}{dx_1}X_1 + \frac{d\Theta}{dx_2}X_2 + \cdots + \frac{d\Theta}{dx_n}X_n\right) dx_1 dx_2 \ldots dx_n.$$

It needs to be noted that the form and meaning of an integral invariant is changed much more significantly when the independent variable called time is changed then when the change of variables only involves the unknown functions $x_1, x_2, \ldots x_n$, because then the laws of motion for the representative point P become completely transformed.

Suppose $n = 3$ and consider x_1, x_2, x_3 as the spatial coordinates of a point P. The equation:

$$\Theta(x_1, x_2, x_3) = 0$$

will represent a surface. Consider an arbitrary portion of this surface and call this portion of surface S.

I will also suppose that at all points on S:

$$\frac{d\Theta}{dx_1}X_1 + \frac{d\Theta}{dx_2}X_2 + \frac{d\Theta}{dx_3}X_3 \neq 0.$$

It results from this that the portion of surface S is not tangent to any trajectory. I will thus state that S is a contactless surface.

Let P_0 be a point on S; a trajectory passes through this point. If the extension of this trajectory again crosses through S at a point P_1, I will state that P_1 is the *recurrence* of P_0. And in turn P_1 can have for recurrence P_2 which I will call the *second recurrence* of P_0 and so on.

If a curve C traced on S is considered, the n recurrences of the various points of this curve will form another curve C' that I will call the nth *recurrence* of C. In the same way, the area would be defined which is the nth *recurrence* of a given area which is part of S.

That stated, let there be a portion of contactless surface S with the equation $\Theta = 0$; let C be a closed curve traced on this surface and delimiting an area A; let C' and A' be the first recurrences, and C^n and A^n be the nth recurrences of C and A.

A trajectory passes through each of these points of C, and I extend this trajectory from its first meeting with C to its meeting with C'. The family of these trajectories will form a trajectory surface T.

I consider the volume V delimited by the trajectory surface T and by the two areas A and A'. Assume that there is a positive invariant:

$$J = \int M \mathrm{d}x_1 \mathrm{d}x_2 \mathrm{d}x_3.$$

I extend this invariant to the volume V and I state that $\mathrm{d}J/\mathrm{d}t$ is zero.

Let $\mathrm{d}\omega$ be an element of the surface S. Follow the normal to this element and on this normal take an infinitesimal length $\mathrm{d}n$. Let $\Theta + \dfrac{\mathrm{d}\Theta}{\mathrm{d}n}\mathrm{d}n$ be the value of Θ at the end of this length. If the normal was followed in the direction of increasing Θ, then:

$$\frac{\mathrm{d}\Theta}{\mathrm{d}n} > 0.$$

Set:

$$\frac{\dfrac{\mathrm{d}\Theta}{\mathrm{d}x_1}X_1 + \dfrac{\mathrm{d}\Theta}{\mathrm{d}x_2}X_2 + \dfrac{\mathrm{d}\Theta}{\mathrm{d}x_3}X_3}{\dfrac{\mathrm{d}\theta}{\mathrm{d}n}} = H,$$

we will then have

$$\frac{\mathrm{d}J}{\mathrm{d}t} = \int_{A'} MH\mathrm{d}\omega - \int_A MH\mathrm{d}\omega,$$

where the first integral is extended to the area A' and the second to the area A.

The integral

$$\int MH\mathrm{d}\omega$$

retains the same value whether it is over the area A, or the area A', or consequently the area A''. It is therefore an integral invariant of a specific kind which retains the same value for an arbitrary area or for one of its recurrences.

These invariants are additionally positive, because by assumption M and H and, as a consequence, MH are positive.

4 Using Integral Invariants

The following theorems are what make integral invariants interesting and we will make frequent use of them.

Above we defined stability by stating that the moving point P must remain at a finite distance; sometimes it will be given a different meaning. For there to be stability, after sufficiently long time the point P has to return if not to its initial position then at least to a position as close to this initial position as desired.

This latter meaning is how Poisson understood stability. When he proved that, if the second powers of the masses are considered, the major axes of the orbits do not change, he only looked at establishing that the series expansion of these major axes only contained periodic terms of the form $\sin \alpha t$ or $\cos \alpha t$ or mixed terms of the form $t \sin \alpha t$ or $t \cos \alpha t$, without including any secular term of the form t or t^2. Which does not mean that the major axes can never exceed a specific value, because a mixed term $t \cos \alpha t$ can grow beyond any limit; it only means that the major axes will go back through their initial value infinitely many times.

Can all the solutions be stable, in the meaning of Poisson? Poisson did not think so, because his proof expressly assumed that the mean motions are not commensurable; the proof therefore does not apply to arbitrary initial conditions of the motion.

The existence of asymptotic solutions, which we will establish later, is sufficient to show that if the initial position of the point P is chosen appropriately, then this point P will not return infinitely many times as close to this initial position as desired.

But I propose to establish that, in one of the specific cases of the three-body problem, the initial position of the point P can be chosen (and can be chosen infinitely many ways) such that this point P returns as close to its initial position as desired infinitely many times.

In other words, there will be infinitely many specific solutions to the problem which will not be stable in the second sense of the word—that is, in the meaning of Poisson; but, there will be infinitely many which are stable. I will add that the first can be regarded as exceptional and later I will seek to understand the precise meaning that I give to this word.

Assume $n = 3$ and consider x_1, x_2, x_3 as the spatial coordinates of a point P.

Theorem I *Assume that the point P remains at a finite distance and that the volume $\int dx_1 dx_2 dx_3$ is an integral invariant; consider an arbitrary region r_0, however, small this region, there will be trajectories which will pass through it infinitely many times.*

In fact, since the point P remains at a finite distance, it will never leave a bounded region R. I call V the volume of this region R.

Now imagine a very small region r_0, I will call the volume of this region v. A trajectory passes through each of the points of r_0; this trajectory can be regarded as the path followed by a point moving according to the law defined by our differential equations. Therefore consider infinitely many moving points which at time zero fill the region r_0 and which then move according to this law. At time τ they will fill some region r_1, at time 2τ a region r_2, etc. and at time $n\tau$ a region r_n. I can assume that τ is large enough and that r_0 is small enough so that r_0 and r_1 have no point in common.

Since the volume is an integral invariant, these various regions $r_0, r_1, \ldots r_n$ will have the same volume v. If these regions have no point in common, then the total

4 Using Integral Invariants

volume would not be larger than nv; on the other hand all these regions are inside R so the total volume is smaller than V. If therefore we have:

$$n > \frac{V}{v},$$

then it must be that at least two of our regions have a common portion. Let r_p and r_q be these two regions ($q > p$). If r_p and r_q have a common portion, it is clear that r_0 and r_{q-p} will have to have a common portion.

More generally, if k regions having a common portion can be found, no point in space could belong to more than $k - 1$ of the regions. The total volume occupied by these regions would therefore be greater than $nv/(k-1)$. If therefore we have:

$$n > (k-1)\frac{V}{v},$$

then it must be possible to find k regions having a common portion. Let:

$$r_{p_1}, r_{p_2}, \ldots r_{p_k}$$

be these regions. Then

$$r_0, r_{p_2-p_1}, r_{p_3-p_1}, \ldots r_{p_k-p_1}$$

will also have a common portion.

But, let us take up the question again from a different perspective. By analogy with the nomenclature from the preceding section, we agree to state that the region r_n is the nth recurrence of r_0, and r_0 is the nth antecedent of r_n.

Suppose then that r_p is the first of the successive recurrences which has a common portion with r_0. Let this common portion be r'_0; let s'_0 be the pth antecedent of r'_0 which would also be part of r_0 because its pth recurrence is part of r_p.

Then let r'_{p_1} be the first of the recurrences of r'_0 which has a common portion with r'_0; let r''_0 be this common portion; its p_1th antecedent will be part of r'_0 and consequently of r_0, and its $p+p_1$th antecedent which I will call s''_0 will be part of s'_0 and consequently of r_0.

Thus s''_0 will be part of r_0 and so will its pth and p_1th recurrences.

And so on.

With r''_0 we will form r'''_0 as we formed r''_0 with r'_0 and r'_0 with r_0; we will then form $r^{IV}_0, \ldots r^n_0, \ldots$.

I will assume that the first of the successive recurrences of r^n_o which has a common portion with r^n_o is that of order p_n.

I will call s^n_0 the antecedent of r^n_o of order $p+p_1+p_2+\ldots p_{n-1}$.

Then s_0^n will be part of r_0 and also of its n recurrences of order:

$$p, p+p_1, p+p_1+p_2, \ldots p+p_1+p_2+\cdots p_{n-1}.$$

Additionally s_0^n will be part of s_0^{n-1}, s_0^{n-1} of s_0^{n-2}, etc.

There will then be points which belong at the same time to the regions $r_0, s_0', s_0'', \ldots s_0^n, s_o^{n+1}, \ldots$ ad infinitum. The set of these points will form a region σ which could additionally reduce to one or several points.

Then the region σ will be part of r_0 and also of its recurrences of order $p, p+p_1$, $p+p_1+p_2, \ldots, p+p_1+p_2+\cdots p_n, p+p_1+p_2+\cdots p_n+p_{n+1}, \ldots$ ad infinitum.

In other words, any trajectory coming from one of the points of σ will traverse the region r_0 infinitely many times.
Which was to be proved.

Corollary *It follows from the preceding that there exist infinitely many trajectories which traverse the region r_0 infinitely many times; but there can exist others which only traverse this region a finite number of times. I now propose to explain why the latter trajectories can be regarded as exceptional.*

Since this expression does not have any precise meaning in itself, I will need to start by filling-in the definition.

We agree to state that the ratio of the probability that the initial position of the moving point P belongs to a certain region r_0 to the probability that this initial position belongs to another region r_0' is equal to the ratio of the volume of r_0 to the volume of r_0'.

With the probabilities thus defined, I propose to establish that the probability is zero that a trajectory coming from a point in r_0 does not traverse this region more than k times, however large k is and however small the region r_0 is. This is what I mean when I state that the trajectories which only traverse r_0 a finite number of times are exceptional.

I assume that the initial position of the point P belongs to r_0 and I propose to calculate the probability that the trajectory coming from this point does not traverse the region r_0 $k+1$ times from the epoch O to the epoch $n\tau$.

We have seen that if the volume v of r_0 is such that:

$$n > \frac{kV}{v},$$

then $k+1$ regions can be found that I will call

$$r_0, r_{\alpha_1}, \ldots r_{\alpha_k}$$

and which will have a common portion. If s_{α_k} is this common portion, let s_0 be its antecedent of order α_k and designate the pth recurrence of s_0 by s_p.

4 Using Integral Invariants

I state that if the initial position of the point P belongs to s_0, then the trajectory coming from this point will cross the region r_0 at least $k+1$ times between the epoch 0 to the epoch $n\tau$.

In fact, the moving point which describes this trajectory will be found in the region s_0 at epoch 0, in the region s_p at epoch $p\tau$, and in the region s_n at the epoch $n\tau$. It will therefore necessarily traverse, between the epochs 0 and $n\tau$, the following regions:

$$s_0, s_{\alpha_k - \alpha_{k-1}}, s_{\alpha_k - \alpha_{k-2}}, \ldots s_{\alpha_k - \alpha_2}, s_{\alpha_k - \alpha_1}, s_{\alpha_k}.$$

Now I state that all these regions are part of r_0. In fact s_{α_k} is part of r_0 by definition; s_0 is part of r_0 because its α_kth recurrence s_{α_k} is part of r_{α_k}, and in general $s_{\alpha_k - \alpha_i}$ is part of r_0 because its α_ith recurrence s_{α_k} is part of r_{α_i}.

Therefore the moving point will pass through the region r_0 at least $k+1$ times. Which was to be proved.

Now let σ_0 be the portion of r_0 that does not belong either to s_0 or to any analogous region, such that the trajectories originating from the various points of σ_0 do not traverse the region r_0 at least $k+1$ times between the epochs 0 and $n\tau$. Let the volume of σ_0 be w.

The probability being sought, meaning the probability that our trajectory does not traverse r_0 $k+1$ times between these two epochs will then be w/v.

Now, by assumption, no trajectory originating from σ_0 traverses r_0, and especially not σ_0, $k+1$ times between these two epochs. We then have:

$$w < \frac{kV}{n}$$

and our probability will be smaller than

$$\frac{kV}{nv}.$$

However large k is and however small v is, n can always be taken large enough such that this expression is as small as we want. Therefore, there is a null probability that our trajectory, which we know originates from a point in r_0, does not traverse this region more than k times since the epoch 0 until the epoch $+\infty$. Which was to be proved.

Extension of theorem I. We assumed that:

(1) $n = 3$
(2) The volume is an integral invariant.
(3) The point P is constrained to remain within a finite distance.

The theorem is still true if the volume is not an integral invariant, provided that there exists an arbitrary positive invariant:

$$\int M dx_1 dx_2 dx_3.$$

It is still true if $n > 3$, if there is a positive invariant:

$$\int M dx_1 dx_2 \cdots dx_n$$

and if $x_1, x_2, \cdots x_n$, which are the coordinates of the point P in the n-dimensional space, are constrained to remain finite.

But there is more.

Suppose that $x_1, x_2, \cdots x_n$ are no longer constrained to remain finite but that the positive integral invariant

$$\int M dx_1 dx_2 \cdots dx_n$$

over the entire n-dimensional space has a finite value. The theorem will still be true.

Here is a case which will come up more frequently.

Assume that an integral of Eq. (1) is known

$$F(x_1, x_2, \cdots x_n) = \text{const.}$$

If $F = \text{const.}$ is the general equation of a family of closed surfaces in n-dimensional space, if, in other words, F is a one-to-one function which becomes infinite when any one of the variables $x_1, x_2, \cdots x_n$ stops being finite, it is clear that $x_1, x_2, \cdots x_n$ will always remain finite, because F keeps a constant finite value; this is therefore within the conditions of the statement of the theorem.

But suppose that the surfaces $F = \text{const.}$ are not closed; it could nonetheless turn out that the positive integral invariant

$$\int M dx_1 dx_2 \cdots dx_n$$

has a finite value over all the families of values of x such that:

$$C_1 < F < C_2;$$

the theorem will again be true.

This is what happens in particular in the following case.

4 Using Integral Invariants

In G. W. Hill's theory of the moon, in a first approximation he neglected the parallax of the sun, the eccentricity of the sun and the inclination of the orbits; he arrived at the following equations:

$$\frac{dx}{dt} = x', \quad \frac{dx'}{dt} = 2n'y' - x\left(\frac{\mu}{\sqrt{(x^2+y^2)^3}} - 3n'^2\right),$$

$$\frac{dy}{dt} = y', \quad \frac{dy'}{dt} = -2n'x' - \frac{\mu y}{\sqrt{(x^2+y^2)^3}},$$

which have the integral

$$F = \frac{x'^2 + y'^2}{2} - \frac{\mu}{\sqrt{x^2+y^2}} - \frac{3}{2}n'^2 x^2 = \text{const.}$$

and the integral invariant

$$\int dx dy dx' dy'.$$

If we regard x, y, x', and y' as the coordinates of a point in four-dimensional space, then the equation $F = \text{const.}$ represents a family of surfaces which are not closed. But the integral invariant over all points included between two of these surfaces is finite, as we will prove.

Theorem I is therefore still true; meaning that there exist trajectories which traverse any region of the four-dimensional space, however small this region might be, infinitely many times.

It remains to calculate the quadruple integral

$$J = \int dx dy dx' dy',$$

where this integral is over all families of values such that

$$C_1 < F < C_2.$$

We change variables and transform our quadruple integral by setting:

$$x' = \cos\varphi\sqrt{2r}, \quad y' = \sin\varphi\sqrt{2r},$$
$$x = \rho\cos\omega, \quad y = \rho\sin\omega;$$

this integral becomes:

$$J = \int \rho d\rho dr d\omega d\varphi$$

and it also follows:

$$F = r - \frac{\mu}{\rho} - \frac{3}{2}n'^2\rho^2\cos^2\omega.$$

We need to first integrate over φ between the limits 0 and 2π, which gives:

$$J = 2\pi \int \rho d\rho dr d\omega$$

and the integration must be over all families of values of ρ, r, and ω which satisfy the inequalities:

$$\begin{aligned} r &> 0, \\ r &> C_1 + \frac{\mu}{\rho} + \frac{3}{2}n'^2\rho^2\cos^2\omega, \\ r &< C_2 + \frac{\mu}{\rho} + \frac{3}{2}n'^2\rho^2\cos^2\omega. \end{aligned} \quad (1)$$

The following can be deduced from these inequalities:

$$C_2 + \frac{\mu}{\rho} + \frac{3}{2}n'^2\rho^2\cos^2\omega > 0.$$

Regard ρ and ω as polar coordinates of a point and construct the curve

$$C_2 + \frac{\mu}{\rho} + \frac{3}{2}n'^2\rho^2\cos^2\omega = 0.$$

We will see that if C_2 is smaller than $-\frac{1}{2}(9n'\mu)^{2/3}$ this curve is composed of a closed oval located entirely inside the circle

$$\rho = \sqrt[3]{\frac{\mu}{3n'^2}}$$

and of two infinite branches located entirely outside the circle.

The reader will be able to do this construction easily; if the reader experiences any difficulty, I suggest they consult the original treatise of G.W. Hill in the *American Journal of Mathematics*, volume 1.

From this G. W. Hill concluded that if the point ρ, ω is inside this closed oval at the beginning of time, it will always remain there and consequently ρ will always remain smaller than $\sqrt[3]{\mu/3n'^2}$. Thus if the parallax of the sun, its eccentricity, and the inclinations are neglected, it will be possible to assign an upper limit to the radius vector of the moon. In fact as it relates to the moon, the constant C_2 is smaller than $-\frac{1}{2}(9n'\mu)^{2/3}$.

4 Using Integral Invariants

I propose to supplement this remarkable result from G. W. Hill by showing that, under these conditions, the moon would also experience stability in the meaning of Poisson; by that I mean that, if the motion's initial conditions are not exceptional, the moon would return as close as one wants to its initial position infinitely many times. That is why, as I explained above, I propose to prove that the integral J is finite.

Since ρ is smaller than $\sqrt[3]{\mu/3n'^2}$ and consequently bounded, the integral:

$$J = 2\pi \int \rho \, d\rho \, dr \, d\omega$$

can only become infinite if r increases indefinitely, and r cannot become infinite in light of the inequalities (1) unless ρ approaches zero.

Therefore set:

$$J = J' + J'',$$

where J' represents the integral over all families of values such that

$$r > 0, \quad \rho > \rho_0, \quad C_1 < F < C_2 \tag{2}$$

and J'' represents the integral over all families of values such that:

$$r > 0, \quad \rho < \rho_0, \quad C_1 < F < C_2. \tag{3}$$

When the inequalities (2) are satisfied ρ cannot become zero; therefore r cannot become infinite. Therefore the first integral, J' is finite.

Now examine J''. I can assume that ρ_0 was taken small enough that

$$C_1 + \frac{\mu}{\rho_0} > 0.$$

The inequalities $F > C_1$ and $\rho < \rho_0$ then lead to $r > 0$. We therefore need to integrate over r between the limits:

$$C_1 + \frac{\mu}{\rho} + \frac{3}{2}n'^2 \rho^2 \cos^2 \omega \quad \text{and} \quad C_2 + \frac{\mu}{\rho} + \frac{3}{2}n'^2 \rho^2 \cos^2 \omega.$$

It then follows:

$$J'' = 2\pi(C_2 - C_1) \int_0^{2\pi} d\omega \int_0^{\rho_0} \rho \, d\rho = 2\pi^2 \rho_0^2 (C_2 - C_1)$$

J'' is therefore finite and consequently also J.
Which was to be proved.

K. Bohlin generalized the result of G. W. Hill in the following way. We consider the following special case of the three-body problem. Let A be a body of mass $1-\mu$, B be a body of mass μ, and C a body of infinitesimal mass. Imagine that the two bodies A and B whose motion must be Keplerian, because it is not perturbed by the mass C, trace out around their mutual center of gravity, assumed to be fixed, two concentric circumferences, and that C moves in the plane of these two circumferences. I will take a constant distance AB as a unit of length, such that the radii of these two circumferences are $1-\mu$ and μ. I will assume that the unit of time has been selected such that the angular speed of the two points A and B on their circumferences is equal to 1 (or that the Gaussian gravitational constant is equal to 1, which amounts to the same thing).

We next select two moving axes with their origin at the center of gravity of the two masses A and B; the first of these axes will be the straight line AB and the second will be perpendicular to the first.

The coordinates of A relative to these two axes are $-\mu$ and 0; those of B are $1-\mu$ and 0; and those for C, I will call x and y; for the equations of motion I then have:

$$\frac{dx}{dt} = x', \quad \frac{dx'}{dt} = 2y' + \frac{dV}{dx} + x,$$
$$\frac{dy}{dt} = y', \quad \frac{dy'}{dt} = -2x' + \frac{dV}{dy} + y,$$

by setting

$$V = \frac{1-\mu}{AC} + \frac{\mu}{BC}.$$

Additionally we have:

$$\overline{AC}^2 = (x+\mu)^2 + y^2, \quad \overline{BC}^2 = (x+\mu-1)^2 + y^2$$

These equations have an integral:

$$F = \frac{x'^2 + y'^2}{2} - V - \frac{x^2 + y^2}{2} = K$$

and an integral invariant:

$$J = \int dx\,dy\,dx'\,dy'$$

K. Bohlin, in *Acta Mathematica* volume 10, generalized the result of G.W. Hill, by showing that if the constant K has a suitable value (which we will assume) and if the initial values of x and y are small enough, these quantities, x and y, will remain bounded.

4 Using Integral Invariants

I now propose to prove that the integral J over all families of values such that

$$K_1 < F < K_2$$

is finite; and from that we will be able to conclude, as we did above, that the stability in the meaning of Poisson pertains again in this case.

If the constants K_1 and K_2 are suitably chosen, the theorem from K. Bohlin shows that x and y will be bounded. As for x' and y', it will not be possible for them to become infinite unless V becomes infinite, meaning if AC approaches zero or if BC approaches zero.

Then set:

$$J = J' + J'' + J''',$$

where the integral J' is over all families of values such that:

$$K_1 < F < K_2, \quad \overline{AC}^2 > \rho_0^2, \quad \overline{BC}^2 > \rho_0^2, \quad \left(\rho_0 < \frac{1}{2}\right)$$

the integral J'' to all families of values such that:

$$K_1 < F < K_2, \quad \overline{AC}^2 < \rho_0^2, \quad \left(\text{hence } \overline{BC}^2 > \rho_0^2\right),$$

and the integral J''' to all families of values such that:

$$K_1 < F < K_2, \quad \overline{BC}^2 < \rho_0^2 \quad \left(\text{hence } \overline{AC}^2 > \rho_0^2\right).$$

Since for none of the families of values over which the integral J' extends do AC or BC become zero, this integral J' is finite.

Now examine the integral J''. I can assume that ρ_0 has been chosen small enough such that:

$$\frac{1-\mu}{\rho_0} + K_1 > 0, \quad \frac{\mu}{\rho_0} + K_1 > 0.$$

In this case $(x'^2 + y'^2)/2$ can vary between the bounds:

$$L_1 = K_1 + \frac{1-\mu}{AC} + \frac{\mu}{BC} + \frac{x^2 + y^2}{2} \quad \text{and} \quad K_2 + \frac{1-\mu}{AC} + \frac{\mu}{BC} + \frac{x^2 + y^2}{2} = L_2,$$

because the smaller of these two bounds is positive.

Then set as above:

$$x' = \sqrt{2r}\cos\varphi, \quad y' = \sqrt{2r}\sin\varphi, \quad \text{hence } r = \frac{x'^2 + y'^2}{2};$$

the integral will become

$$J'' = \int dx\,dy\,dr\,d\varphi$$

and it will be necessary to integrate over φ between the bounds 0 and 2π, and over r between the bounds L_1 and L_2; it will then become:

$$J'' = 2\pi(K_2 - K_1)\int dx\,dy.$$

The double integral $\int dx\,dy$ will then need to be over all families of values such that $\overline{AC}^2 < \rho_0^2$; it is therefore equal to $\pi\rho_0^2$; such that it becomes:

$$J'' = 2\pi^2\rho_0^2(K_2 - K_1).$$

J'' is therefore finite, and so are J''' and J.
Which was to be proved.

We therefore have to conclude that (if the initial conditions of motion are not exceptional in the meaning given to this word in the corollary to theorem I) the third body C will go back as close as one wants to its initial position infinitely many times.

In the general case of the three-body problem, it can no longer be affirmed that it will still be the same.

Theorem II *If $n = 3$ and x_1, x_2, x_3 represent the coordinates of a point in ordinary space, and if there is a positive invariant, there cannot be a closed contactless surface.*

In fact let

$$J = \int M\,dx_1\,dx_2\,dx_3$$

be a positive integral invariant. Assume that there is a closed and contactless surface, having the equation

$$F(x_1, x_2, x_3) = 0.$$

Let V be the volume delimited by this surface; we extend the invariant J to this entire volume.

4 Using Integral Invariants

Since the surface S is contactless, the expression:

$$\frac{dF}{dx_1}X_1 + \frac{dF}{dx_2}X_2 + \frac{dF}{dx_3}X_3$$

cannot become zero and consequently change sign; to be concrete, we will assume that it is positive.

Let $d\omega$ be a differential element of the surface S; take the normal to this element from the side of increasing F; take on this normal an infinitesimal segment dn. Let $\frac{dF}{dn}dn$ be the value of F at the end of this segment. We will then have:

$$\frac{dF}{dn} > 0.$$

since J is an invariant, we should have

$$\frac{dJ}{dt} = 0.$$

But we find

$$\frac{dJ}{dt} = \int M \frac{\dfrac{dF}{dx_1}X_1 + \dfrac{dF}{dx_2}X_2 + \dfrac{dF}{dx_3}X_3}{\dfrac{dF}{dn}} d\omega.$$

The integral on the right-hand side, over the entire surface S, is positive because the function within the integral sign is always positive.

We have arrived therefore at two contradictory results and we have to conclude that a closed contactless surface cannot exist.

Extension of Theorem II. It is easy to extend this theorem to the case of $n > 3$; to do that it is sufficient to translate it into analytical language, because geometric representation is no longer possible, and state:

If there is a positive integral invariant, there cannot exist a one-to-one function $F(x_1, x_2, \cdots x_n)$ which is positive, which becomes infinite each time one of the x stops being finite and which is such that

$$\frac{dF}{dt} = \frac{dF}{dx_1}X_1 + \frac{dF}{dx_2}X_2 + \cdots \frac{dF}{dx_n}X_n$$

always has the same sign when F is zero.

To make the importance of this theorem understood, I will limit myself to observing that it is a generalization of the one which I used for proving the legitimacy of Lindstedt's beautiful method.

However, with a perspective to subsequent applications, I prefer to give it a little bit different form by introducing into it a new concept: that of invariant curves.

At the end of the previous section we had considered a portion of surface S, defined by the equation

$$\Theta(x_1, x_2, x_3) = 0$$

and such that for all points of S it holds that

$$\frac{d\Theta}{dx_1}X_1 + \frac{d\Theta}{dx_2}X_2 + \frac{d\Theta}{dx_3}X_3 > 0,$$

such that S is a portion of contactless surface.

We have subsequently defined what was to be understood by the nth recurrence of a point from S or by the nth recurrence of a curve or an area belonging to S. I now understand and from now on I will understand the word recurrence in the meaning of the previous section and not in the meaning used above in the proof of Theorem I.

We have seen that if there is a positive invariant

$$\iiint M dx_1 dx_2 dx_3,$$

there is also another integral

$$\int MH d\omega$$

which must be over all the elements $d\omega$ of an area belonging to S and which has the following properties:

(1) The quantity under the integral sign, MH, is always positive.
(2) The integral has the same value for an arbitrary area belonging to S and for all areas of its recurrences which exist.

With that stated, I will call nth order *invariant curve* any curve traced on S and which coincides with its nth recurrence.

In most questions from dynamics, some very small parameters enter such that one is naturally led to develop solutions following increasing powers of these parameters. Such are the masses in celestial mechanics.

We will therefore imagine that our differential equations

$$\frac{dx_1}{dt} = X_1, \quad \frac{dx_2}{dt} = X_2, \quad \frac{dx_3}{dt} = X_3$$

depend on a parameter μ. We will suppose that X_1, X_2, X_3 are given functions of x_1, x_2, x_3 and μ which could be expanded in increasing powers of μ and that μ is very small.

4 Using Integral Invariants

Now consider an arbitrary function of μ; I assume that this function approaches 0 when μ approaches 0, such that the ratio of this function to μ^n approaches a finite limit. I will state that this function of μ is a very small quantity of nth order.

It needs to be indicated that it is not necessary for it to be possible to expand this function of μ in powers of μ.

With that established, let A_0 and B_0 be two points on a contactless surface S, and let A_1 and B_1 be their recurrences. If the position of A_0 and B_0 depends on μ according to an arbitrary law, then so will the position of A_1 and B_1. I am proposing to prove the following lemmas:

Lemma I *If a portion of contactless surface S passing through the point a_0, b_0, c_0 is considered; if x_0, y_0, z_0 are coordinates of a point on S and if x_1, y_1, z_1 are coordinates of its recurrence, then x_1, y_1, z_1 are expandable in powers of $x_0 - a_0, y_0 - b_0, z_0 - c_0$ and μ provided that these quantities are sufficiently small.*

I can always take for origin the point a_0, b_0, c_0 such that

$$a_0 = b_0 = c_0 = 0.$$

If then

$$z = \varphi(x, y)$$

is the equation of the surface S; this surface will pass through the origin O and one will have:

$$\varphi(0, 0) = 0.$$

I will additionally assume that the function $\varphi(x, y)$ is mapping at all the points on the portion of surface S considered. One trajectory passes through the origin O; imagine that when $\mu = 0$ this trajectory crosses the surface S at time $t = \tau$ at a point P whose coordinates will be:

$$x = a, \quad y = b, \quad z = c$$

According to the terminology that we have adopted, *when it is assumed that* $\mu = 0$, the point P will be the recurrence of the point O.

Now let x_0, y_0, z_0 be a point A very close to O and belonging to the surface S. If a trajectory passes through this point A, and if it is assumed that μ stops being zero but remains very small, it will be seen that this trajectory will come, at an epoch t only slightly different from τ, to cross the surface S at a point B very near P.

This point B, whose coordinates I will call x_1, y_1, z_1, will according to our terminology be the recurrence to the point A.

What I propose to prove is that x_1, y_1, z_1 are expandable in increasing powers of x_0, y_0, z_0 and μ.

In fact according to Theorem III from Sect. 2 of Chap. 1, if x, y, z are coordinates at time t of the moving point which describes the trajectory coming from point A and if additionally x_0, y_0, z_0, μ and $t - \tau$ are sufficiently small, then one will have:

$$\begin{aligned} x &= \psi_1(t - \tau, \mu, x_0, y_0, z_0), \\ y &= \psi_2(t - \tau, \mu, x_0, y_0, z_0), \\ z &= \psi_3(t - \tau, \mu, x_0, y_0, z_0), \end{aligned} \qquad (4)$$

where ψ_1, ψ_2 and ψ_3 are series ordered in powers of $t - \tau, \mu, x_0, y_0$ and z_0.

These series will reduce, respectively, to a, b, c for

$$t - \tau = \mu = x_0 = y_0 = z_0 = 0.$$

Since $\varphi(x, y)$ is expandable in powers of $x - a$ and $y - b$, if $x - a$ and $y - b$ are small enough, we will also have:

$$\varphi(x, y) = \psi_4(t - \tau, \mu, x_0, y_0, z_0),$$

where ψ_4 is a series with the same form as ψ_1, ψ_2 and ψ_3.

We write that the point x, y, z is located on the surface S; we will have:

$$\psi_3 = \psi_4 \qquad (5)$$

The relation (5) can be regarded as a relation between $t - \tau, \mu, x_0, y_0$ and z_0, and one can try to solve it for $t - \tau$.

For:

$$t - \tau = \mu = x_0 = y_0 = z_0 = 0$$

this relation is satisfied because one has

$$\psi_3 = \psi_4 = 0.$$

According to a theorem by Cauchy, which we proved in one of the preceding sections, one can draw $t - \tau$ from the relationship (5) in the following form:

$$t - \tau = \theta(\mu, x_0, y_0, z_0), \qquad (6)$$

where θ is a series ordered in powers of μ, x_0, y_0 and z_0.

The only exception would be if for

$$t - \tau = \mu = x_0 = y_0 = z_0 = 0$$

4 Using Integral Invariants

it held that:

$$\frac{d\psi_3}{dt} = \frac{d\psi_4}{dt}.$$

Now this equation expresses that the trajectory starting from point O for $\mu = 0$ is going to *touch* the surface S at point P.

But it can not be that way, because we will always assume that S is a contactless surface or a portion of contactless surface.

In Eq. (4) we replace $t - \tau$ by θ and x, y, z by x_1, y_1, z_1; it follows:

$$x_1 = \Theta_1(\mu, x_0, y_0, z_0),$$
$$y_1 = \Theta_2(\mu, x_0, y_0, z_0),$$
$$z_1 = \Theta_3(\mu, x_0, y_0, z_0),$$

where Θ_1, Θ_2 and Θ_3 are expanded in powers of μ, x_0, y_0 and z_0.
Which was to be proved.

Lemma II *If the distance between two points A_0 and B_0 belonging to a portion of the contactless surface S is a very small quantity of order n, then so will the distance between their recurrences A_1 and B_1.*

In fact, let a_1, a_2, a_3 be the coordinates of a fixed point P_0 from S very near A_0 and B_0; and let a'_1, a'_2, a'_3 be the coordinates of its recurrence P_1.

Let x_1, x_2, x_3; x'_1, x'_2, x'_3; y_1, y_2, y_3; and y'_1, y'_2, y'_3 be the coordinates of A_0, A_1, B_0 and B_1.

According to Lemma I $x'_1 - a'_1, x'_2 - a'_2, x'_3 - a'_3$ are expandable in increasing powers of $x_1 - a_1, x_2 - a_2, x_3 - a_3$ and μ.

The expression for $y'_1 - a'_1, y'_2 - a'_2, y'_3 - a'_3$ as a function of $y_1 - a_1, y_2 - a_2, y_3 - a_3$ and μ will obviously be the same as that for $x'_1 - a'_1, x'_2 - a'_2, x'_3 - a'_3$ as a function of $x_1 - a_1, x_2 - a_2, x_3 - a_3$ and μ.

From that we conclude that it is possible to write:

$$x'_1 - y'_1 = (x_1 - y_1)F_1 + (x_2 - y_2)F_2 + (x_3 - y_3)F_3,$$
$$x'_2 - y'_2 = (x_1 - y_1)F'_1 + (x_2 - y_2)F'_2 + (x_3 - y_3)F'_3, \quad (7)$$
$$x'_3 - y'_3 = (x_1 - y_1)F''_1 + (x_2 - y_2)F''_2 + (x_3 - y_3)F''_3,$$

where the F are series expanded in powers of:

$$\mu, x_1 - a_1, x_2 - a_2, x_3 - a_3, y_1 - a_1, y_2 - a_2, y_3 - a_3.$$

The quantities $F_1, F_2,$, etc. are finite; therefore if $x_1 - y_1, x_2 - y_2$, and $x_3 - y_3$ are very small quantities of order n, then $x'_1 - y'_1, x'_2 - y'_2$, and $x'_3 - y'_3$ will be also.
Which was to be proved.

Theorem III Let A_1AMB_1B be an invariant curve, such that A_1 and B_1 are the recurrences of A and B. I assume that the arcs AA_1 and BB_1 are very small (approach 0 with μ) but that their curvature is finite.

I assume that this invariant curve and the position of points A and B depend on μ according to an arbitrary law. I assume that there exists a positive integral invariant. If the distance AB is very small of nth order, and the distance AA_1 is not very small of nth order, the arc AA_1 crosses the arc BB_1.

I can always join the points A and B by curve AB located entirely on the portion of contactless surface S and for which the total length is the same order of magnitude as the distance AB meaning a very small quantity of nth order. Let A_1B_1 be an arc of curve which is the recurrence of AB, it will thus be very small of nth order according to Lemma II.

Now here are the various scenarios that are conceivable:

Scenario 1. The two arcs AA_1 and BB_1 cross. I propose to establish that this is the actual scenario.
Scenario 2. The quadrilateral AA_1B_1B is such that the four arcs which are its sides have no other point in common than the four corners A, A_1, B, and B_1. This is the case from Fig. 1.
Scenario 3. The two arcs AB and A_1B_1 cross. This is the case from Fig. 2.
Scenario 4. One of the arcs AB or A_1B_1 crosses one of the arcs AA_1 or BB_1; but the arcs AA_1 and BB_1 do not cross nor do the two arcs AB and A_1B_1.

If there is a positive invariant, then according to the preceding section there will exist a certain integral

$$\int MH d\omega$$

all the elements of which will be positive and which will have to have the same value for the area ABB_1MA and for its recurrence AA_1B_1MA.

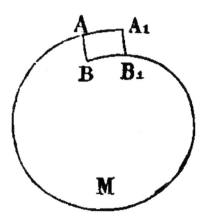

Fig. 1 The corners are the only common points of the four arcs

Fig. 2 The two arcs AB and A_1B_1 cross

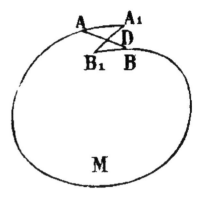

This integral over the area

$$ABA_1B_1 = AA_1B_1MA - ABB_1MA$$

must therefore be zero and as all the elements of the integral are positive, the arrangement cannot be that from Fig. 1 where the area ABA_1B_1 is convex.

The second scenario must therefore be rejected.

In fact in the triangle ADA_1, the distances AD and A_1D are very small of nth order because they are smaller than the arcs AD and A_1D, which are smaller than the arcs AB or A_1B_1 which are of nth order. Furthermore it holds that:

$$AA_1 < AD + A_1D.$$

The distance AA_1 would therefore be a very small quantity of nth order which is contrary to the statement of the theorem.

The third scenario must therefore be rejected.

I state that the fourth scenario cannot be accepted either. Assume in fact for example that the arc AB crosses the arc AA_1 at a point A'. Let ANA' be the portion of the arc AB which goes from A to A'; let APA' be the portion of arc AA_1 which goes from A to A'.

I state that the arc $ANA'B$ can be replaced by the arc $APA'B$; and that the new arc $APA'B$ will be a very small quantity of nth order like the primitive arc $ANA'B$.

In fact the arc ANA' is smaller than AB and it is therefore of nth order; the distance AA' is therefore itself of nth order; the arc APA' is smaller than AA_1 which is very small—meaning it approaches 0 with μ; the arc APA' is therefore very small and its curvature is finite; therefore a bound can be assigned to the ratio of the arc APA' to its chord AA'; this ratio is finite and AA' is of nth order; therefore APA' is of nth order, which was to be proved.

Furthermore the new arc $APA'B$ no longer crosses the arc AA_1 and it only has a common portion APA' with it.

This falls back on the second scenario which was already rejected.

The first scenario is therefore the only one acceptable and the theorem is therefore proved.

Remark In the statement of the theorem we have assumed that the arcs AA_1 and BB_1 are very small and that there curvature is finite. In reality we have only made use of this assumption for showing that if the chord AA' is very small of nth order, it is the same for the arc APA'.

The theorem will therefore still be true even if the arc AA_1 is no longer very small and its curvature finite, provided that it is possible to assign an upper bound to the ratio of an arbitrary arc (which is part of AA_1 or BB_1) to its chord.

Chapter 3
Theory of Periodic Solutions

1 Existence of Periodic Solutions

Consider a system of differential equations

$$\frac{dx_i}{dt} = X_i \quad (i = 1, 2, \ldots n), \tag{1}$$

where the X are functions of x and a parameter μ. The X can also depend on t, but they will then be periodic functions of this variable and the period will be 2π.

Assume that when the parameter μ has the value 0, these equations allow a periodic solution, such that

$$x_i = \varphi_i(t),$$

where φ_i is a periodic function of time whose period will be for example 2π.

Set

$$x_i = \varphi_i + \xi_i$$

and for very small values of μ seek to find values of ξ that we also assume to be very small; it follows that

$$\frac{d\xi_i}{dt} = \mu \frac{dX_i}{d\mu} + \sum_k \xi_k \frac{dX_i}{dx_k}.$$

In the partial derivatives of X, the x_i are replaced by the periodic functions φ_i. The ξ are thus determined by the linear equations with a right-hand side whose coefficients are periodic functions.

Two cases can come up.

(1) The equations without a right-hand side

$$\frac{d\xi_i}{dt} = \sum \xi_k \frac{dX_i}{dx_k} \tag{2}$$

do not allow a periodic solution with period 2π.

In this case, the equations with a right-hand side have a solution which I will write as

$$\xi_i = \mu \psi_i(t),$$

where ψ is a periodic function with period 2π.

(2) The equations without a right-hand side allow a periodic solution with period 2π.

Then it is possible that the equations with a right-hand side do not have a periodic solution, such that in general we will find a solution of the following form:

$$\xi_i = \mu t \psi_{1,i}(t) + \mu \psi_{0,i}(t),$$

where the ψ are still periodic functions, or even in some cases:

$$\xi_i = \mu \left[t^n \psi_{n,i}(t) + t^{n-1} \psi_{n-1,i}(t) + \cdots \psi_{0,i}(t) \right].$$

Now put ourselves in the first case and look at it more closely.

Let us look at forming a periodic solution and expanding it in powers of μ; consequently set

$$x_i = \varphi_i + \mu \varphi_{1,i} + \mu^2 \varphi_{2,i} + \cdots.$$

When these values are substituted for the x_i in the X_i, it will be found that

$$X_i = X_{o,i} + \mu X_{1,i} + \mu^2 X_{2,i} + \cdots.$$

It is clear that the $X_{o,i}$ depend only on the φ_i; the $X_{1,i}$ on the φ_i and the $\varphi_{1,i}$; the $X_{2,i}$ on the $\varphi_{1,i}$ and the $\varphi_{2,i}$; etc. Furthermore, if the $\varphi_{n,i}$ are periodic functions of t with period 2π, then the $X_{n,i}$ will be too.

Additionally, we have

$$X_{n,i} = \sum_k \frac{dX_i}{dx_k} \varphi_{n,k} + Y_{n,i}.$$

1 Existence of Periodic Solutions

In the right-hand side, in the derivatives dX_i/dx_k, the φ_i must be substituted in place of the x_i, as we did above. Additionally, $Y_{n,i}$ will only depend on the φ_i, the $\varphi_{1,i}$, the $\varphi_{2,i}, \cdots$, and the $\varphi_{n-1,i}$, but will no longer depend on the $\varphi_{n,i}$.

With that set, we are led to write the following:

$$\frac{d\varphi_{n,i}}{dt} = \sum_x \frac{dX_i}{dx_k}\varphi_{n,k} + Y_{n,i}. \tag{3}$$

Assume that the quantities

$$\varphi_{1,i}, \varphi_{2,i}, \ldots \varphi_{n-1,i}$$

have been determined using the preceding equations in the form of periodic functions of t; it would then be possible using Eq. (3) to determine the $\varphi_{n,i}$.

Equation (3) is the linear equation in the second term and the coefficients are periodic.

By assumption the equations without a second term

$$\frac{d\varphi_{n,i}}{dt} = \sum_k \frac{dX_i}{dx_k}\varphi_{n,k},$$

which are nothing other than Eq. (2), do not have a periodic solution; therefore, Eq. (3) has one.

From that it results that there exist series

$$x_i = \phi_i + \mu\varphi_{1,i} + \mu^2\varphi_{2,i} + \cdots$$

whose coefficients are periodic and that formally satisfy Eq. (1).

The convergence of the series remains to be proved. There is no doubt that this proof could be done directly; I am not going to do it, however, because, by taking up the question from a different perspective, I am going to rigorously prove the existence of periodic solutions, which leads to the convergence of our series. We will actually only need to rely on the best-known principles from the "calculation of limits."

Let $\phi_i(0) + \beta_i$ be the value of x_i for $t = 0$. Let $\phi_i(0) + \gamma_i$ be the value of x_i for $t = 2\pi$. The γ_i will obviously depend on μ and the initial values of variables and they will become zero with them.

This leads me to write

$$\gamma_i = \beta_i + \alpha_i\mu + \sum b_{ik}\beta_k + \sum [m, p_1, p_2, \ldots p_n]\mu^m \beta_1^{p_1} \beta_2^{p_2} \ldots \beta_n^{p_n} = \beta_i + \psi_i,$$

where the a, b, and the $[m, p_1, p_2, \ldots p_n]$ are constant coefficients.

Periodic solutions with period 2π can be obtained by looking for the cases, where

$$\gamma_i = \beta_i.$$

Therefore, μ can be considered as a given of the problem and one can look to solve the equations

$$\psi_1 = \psi_2 = \cdots \psi_n = 0 \tag{4}$$

relative to the n unknowns β. We know that the ψ are mapping functions of μ and β that become zero with the variables (See Theorem III, Sect. 2 in Chap. 1.)

If the Jacobian determinant of the ψ relative to the β (meaning the determinant of the b_{ik}) is not zero, then these n equations can be solved and the solution

$$\beta_i = \theta_i(\mu)$$

found, where the θ_i are, according to a well-known theorem, mapping functions of μ approaching zero with μ. (See Theorem IV in Sect. 2 in Chap. 1.)

This is the case that we studied above and where Eq. (2) does not have a periodic solution.

From that it must be concluded that for sufficiently small values of μ, Eq. (1) does not allow a periodic solution.

We now assume that the Jacobian determinant of the ψ is zero; in virtue of Theorem VI in Sect. 2 in Chap. 1, we will then be able to eliminate the $\beta_1, \beta_2, \ldots \beta_{n-1}$ from Eq. (4); we will thus arrive at a single equation

$$\Phi = 0$$

whose left-hand side will be expanded in powers of μ and β_n.

There would only be an exception if Eq. (4) were not distinct; but in that case we will add another arbitrarily chosen equation to them.

If we regard μ and β_n as coordinates of a point in a plane, then the equation $\Phi = 0$ represents a curve passing through the origin. A periodic solution will correspond to each of the points of this curve, such that in order to study the periodic solutions which correspond to the small values of μ, and the β, it will be sufficient for us to construct the part of this curve neighboring the origin.

If the Jacobian determinant of the ψ is zero, then we will have (for $\mu = \beta_n = 0$)

$$\frac{d\Phi}{d\beta_n} = 0.$$

In other words, the curve $\Phi = 0$ will be tangent to the origin at the straight-line $\mu = 0$, or else again for $\mu = 0$, the equation $\Phi = 0$ will be an equation in β_n which will have 0 as a multiple root; I will call m the order of the multiplicity of this root.

1 Existence of Periodic Solutions

In virtue of Theorem V in Sect. 2 in Chap. 1, m series expandable in fractional, positive powers of μ can be found that become zero with μ and which satisfy the equation $\Phi = 0$ when substituted in place of β_n.

Let us consider the intersection of the curve $\Phi = 0$, or rather the portion of this curve near the origin, with two straight lines $\mu = \varepsilon$ and $\mu = -\varepsilon$ very close to the straight-line $\mu = 0$. The points of intersection can be obtained by setting $\mu = \varepsilon$ and then $\mu = -\varepsilon$ in the m series which I just spoke about.

Let m_1 be the number of points of intersection of $\Phi = 0$ and $\mu = +\varepsilon$ that are real and near the origin. Let m_2 be the number of points of intersection of $\Phi = 0$ and $\mu = -\varepsilon$ that are real and near the origin.

The three numbers m, m_1, and m_2 will have the same parity.

Therefore, if m is odd, m_1 and m_2 will be at least equal to 1. Therefore, there exist periodic solutions for the smallest values of μ, both positive and negative.

How can a periodic solution disappear when μ is varied continuously? How is it possible that the number of solutions for $\mu = +\varepsilon$ be smaller than for $\mu = -\varepsilon$, that $m_1 < m_2$?

I first observe that a periodic solution can only disappear when μ goes from the value $-\epsilon$ to the value $+\varepsilon$ if for $\mu = 0$, the equation $\Phi = 0$ has a multiple root; in other words *a periodic solution can only disappear after merging with another periodic solution*. Additionally, since m_1 and m_2 have the same parity the difference $m_2 - m_1$ is always even.

Thus, periodic solutions disappear in pairs in the same way as the real roots of algebraic equations.

An interesting specific case is the one where for $\mu = 0$, the differential Eq. (1) have infinitely many periodic solutions that I will write as

$$x_1 = \varphi_1(t, h), \quad x_2 = \varphi_2(t, h), \quad \ldots \quad x_n = \varphi_n(t, h),$$

where h is an arbitrary constant.

In this case, Eq. (4) is no longer distinct for $\mu = 0$, and Φ contains μ as a factor such that we are able to set

$$\Phi = \mu \Phi_1,$$

where Φ_1 is a mapping function of β_n and μ; additionally, Φ_1 also depends on h. The curve $\Phi = 0$ is then broken down into two others, specifically the straight-line $\mu = 0$ and the curve $\Phi_1 = 0$; this latter curve is the one that needs to be studied.

The curve $\Phi = 0$ necessarily passes through the origin; it is not always the same for $\Phi_1 = 0$; first it is necessary to arrange for it pass through there, by suitable choice of h. Once it is been made to pass through there, it will be studied as was done for the curve $\Phi = 0$.

If for $\mu = \beta_n = 0$, $d\Phi_1/d\beta_n$ is not zero (or, more generally, if for $\mu = 0$ the equation $\Phi_1 = 0$ allows $\beta_n = 0$ as an odd-order multiple root), then there will still be periodic solutions for small values of μ.

It will often happen that, even before the elimination, some of the functions ψ_i contain μ as a factor. In this case, one would start by dividing the corresponding equations by μ.

If Eq. (1) allows a one-to-one integral,

$$F(x_1, x_2, \ldots x_n) = \text{const.}$$

Equation (4) will not be distinct unless

$$\frac{dF}{dx_1} = \frac{dF}{dx_2} = \cdots \frac{dF}{dx_n} = 0$$

holds simultaneously for

$$x_1 = \varphi_1(0), \quad x_2 = \varphi_2(0), \quad \cdots \quad x_n = \varphi_n(0).$$

In fact, it will follow identically

$$F[\varphi_i(0) + \beta_i + \psi_i] = F[\varphi_i(0) + \beta_i].$$

If for example for $x_i = \varphi_i(0)$, dF/dx_i is not zero; this equation could be drawn:

$$\psi_1 = \psi_2 \Theta_2 + \psi_3 \Theta_3 + \cdots \psi_n \Theta_n,$$

where $\Theta_2, \Theta_3, \cdots \Theta_n$ are series arranged in increasing powers of

$$\beta_1, \beta_2, \ldots \beta_n, \psi_2, \psi_3 \ldots \psi_n.$$

The first of Eq. (4) is therefore a consequence of the last $n - 1$. It will then be deleted to be replaced by another arbitrarily chosen equation.

In the preceding, we have assumed that the functions $X_1, X_2, \ldots X_n$ that enter into the differential Eq. (1) depend on time t. The results would be modified if the time t does not enter into these equations.

First, there is a difference between the two cases which is impossible to miss. In the preceding, we had assumed that the X_i would periodic functions of time and that the period was 2π; from this it followed that, if the equations allowed periodic solution, then the period of this solution would be equal to 2π or a multiple of 2π. If, on the other hand, the X_i were independent of t, then the period of the periodic solution could be arbitrary.

Second, if Eq. (1) allows a periodic solution (and if the X do not depend on t), then they allow infinitely many of them.

If in fact

$$x_1 = \varphi_1(t), \quad x_2 = \varphi_2(t), \quad \cdots \quad x_n = \varphi_n(t),$$

1 Existence of Periodic Solutions

is a periodic solution of Eq. (1), then

$$x_1 = \varphi_1(t+h), \quad x_2 = \varphi_2(t+h), \quad \ldots \quad x_n = \varphi_n(t+h)$$

will also be periodic (whatever the constant h).

Thus the case, which we first covered and in which for $\mu = 0$ Eq. (1) allows one and only one periodic solution, can only occur if the X do not depend on t

Therefore, we work with the case where the time t does not explicitly enter into Eq. (1) and we assume that for $\mu = 0$, these equations allow a periodic solution of period T:

$$x_1 = \varphi_1(t), \quad x_2 = \varphi_2(t), \quad \ldots \quad x_n = \varphi_n(t).$$

Let $\varphi_i(0) + \beta_i$ be the value of x_i for $t = 0$; let $\varphi_i(0) + \gamma_i$ be the value of x_i for $t = T + \tau$. Next set, as we did above,

$$\gamma_i - \beta_i = \psi_i$$

The ψ_i will be mapping functions of μ, of $\beta_1, \beta_2, \ldots \beta_n$, and of τ which approach 0 when these variables do.

We, therefore, need to solve the n equations

$$\psi_1 = \psi_2 = \cdots \psi_n = 0 \tag{5}$$

with respect to the $n + 1$ unknowns

$$\beta_1, \beta_2, \ldots \beta_n, \tau.$$

There is, therefore, one unknown too many and we can, for example, arbitrarily set

$$\beta_n = 0.$$

From Eq. (5), we derive $\beta_1, \beta_2, \ldots \beta_n$ and τ as mapping functions of μ that are zero when μ is. This is possible as long as the determinant:

$$\begin{vmatrix} \dfrac{d\psi_1}{d\beta_1} & \dfrac{d\psi_1}{d\beta_2} & \cdots & \dfrac{d\psi_1}{d\beta_{n-1}} & \dfrac{d\psi_1}{d\tau} \\ \dfrac{d\psi_2}{d\beta_1} & \dfrac{d\psi_2}{d\beta_2} & \cdots & \dfrac{d\psi_2}{d\beta_{n-1}} & \dfrac{d\psi_2}{d\tau} \\ \vdots & \vdots & \ddots & \vdots & \vdots \\ \dfrac{d\psi_n}{d\beta_1} & \dfrac{d\psi_n}{d\beta_2} & \cdots & \dfrac{d\psi_n}{d\beta_{n-1}} & \dfrac{d\psi_n}{d\tau} \end{vmatrix}$$

is only zero for $\mu = \beta_i = \tau = 0$.

If this determinant were zero, instead of arbitrarily setting $\beta_n = 0$, one could for example set $\beta_i = 0$, and the method would only fail if all the determinants in the matrix:

$$\begin{Vmatrix} \dfrac{d\psi_1}{d\beta_1} & \dfrac{d\psi_1}{d\beta_2} & \cdots & \dfrac{d\psi_1}{d\beta_n} & \dfrac{d\psi_1}{d\tau} \\ \dfrac{d\psi_2}{d\beta_1} & \dfrac{d\psi_2}{d\beta_2} & \cdots & \dfrac{d\psi_2}{d\beta_n} & \dfrac{d\psi_2}{d\tau} \\ \vdots & \vdots & \ddots & \vdots & \vdots \\ \dfrac{d\psi_n}{d\beta_1} & \dfrac{d\psi_n}{d\beta_2} & \cdots & \dfrac{d\psi_n}{d\beta_n} & \dfrac{d\psi_n}{d\tau} \end{Vmatrix}$$

were zero at the same time. (It should be remarked that the determinant obtained by eliminating the last column from this matrix is always zero for $\mu = \beta_i = \tau = 0$.)

Since, in general all these determinants will not be zero at the same time, Eq. (1) will for small values of μ allow a periodic solution with period $T + \tau$.

2 Characteristic Exponents

Return to the equations

$$\frac{dx_i}{dt} = X_i \qquad (1)$$

and imagine that they allow a periodic solution

$$x_i = \varphi_i(t).$$

Form the perturbation equations (see Chap. 1) of Eq. (1) by setting

$$x_i = \varphi_i(t) + \xi_i$$

and neglecting the squares of the ξ.

These series expansions will be written as

$$\frac{d\xi_i}{dt} = \frac{dX_i}{dx_1}\xi_1 + \frac{dX_i}{dx_2}\xi_2 + \cdots \frac{dX_i}{dx_n}\xi_n. \qquad (2)$$

These equations are linear in the ξ, and their coefficients dX_i/dx_k (when x_i is replaced by $\varphi_i(t)$ in them) are periodic functions of t. We, therefore, need to integrate these linear equations with periodic coefficients.

2 Characteristic Exponents

The form of the solutions of these equations is known in general; n specific solutions of the following form are obtained as follows:

$$\begin{aligned}
\xi_1 &= e^{a_1 t} S_{11} & \xi_2 &= e^{a_1 t} S_{21} & \cdots & & \xi_n &= e^{a_1 t} S_{n1} \\
\xi_1 &= e^{a_2 t} S_{12} & \xi_2 &= e^{a_2 t} S_{22} & \cdots & & \xi_n &= e^{a_2 t} S_{n2} \\
&\vdots & & \vdots & \ddots & & &\vdots \\
\xi_1 &= e^{a_n t} S_{1n} & \xi_2 &= e^{a_n t} S_{2n} & \cdots & & \xi_n &= e^{a_n t} S_{nn},
\end{aligned} \qquad (3)$$

where the α are constants and the S_{ik} are periodic functions of t with the same period as the $\varphi_i(t)$.

The constants α are called the *characteristic exponents* of the periodic solution.

If α is purely imaginary such that its square is negative, then the magnitude of $e^{\alpha t}$ is constant and equal to 1. If, on the other hand, α is real, or if α is complex such that its square is not real, then the magnitude of $e^{\alpha t}$ approaches infinity for $t = +\infty$ or for $t = -\infty$. However, if the squares of all the α are real and negative, then the quantities $\xi_1, \xi_2, \ldots \xi_n$ will remain finite; in the other case, I will state that this solution is unstable.

An interesting special case is the one where two or more of the characteristic exponents are equal to each other. In this case the solutions of Eq. (2) can no longer be put into form (3). If for example

$$\alpha_1 = \alpha_2,$$

Eq. (2) will allow two specific solutions which would be written as

$$\xi_i = e^{\alpha_1 t} S_{i,1}$$

and

$$\xi_i = t e^{\alpha_1 t} S_{i,1} + e^{\alpha_1 t} S_{i,2},$$

where the $S_{i,1}$ and the $S_{i,2}$ are periodic functions of t.

If three of the characteristic exponents are equal to each other, then not only t but also t^2 will be seen to appear outside of the trigonometric and exponential functions.

Assume that time t does not explicitly enter into Eq. (1) such that the functions X_i do not depend on this variable; additionally assume that Eq. (1) allows an integral

$$F(x_1, x_2, \ldots x_n) = C. \qquad (4)$$

It is easy to see that in this case two of the characteristic exponents are zero.

This case is then part of the exceptional case that we just reported; but no difficulty comes from it; it is in fact easy using the integral (4) to lower the order of Eq. (1) by one. There are then only $n - 1$ characteristic exponents and there is only one which is zero.

We are now going to consider a specific case which is the one where Eq. (1) has the form of the equations of dynamics. Therefore, we write them in the following form:

$$\frac{dx_i}{dt} = \frac{dF}{dy_i}, \quad \frac{dy_i}{dt} = -\frac{dF}{dx_i}, \quad (i = 1, 2, \ldots n), \tag{1'}$$

where F is an arbitrary function of $x_1, x_2, \ldots x_n, y_1, y_2, \ldots y_n$; we can assume that F is independent of t; or that F depends not only on the x and the y, but also on t, and that with respect to this last variable, it is a periodic function with period 2π.

Assume that Eq. (1') allows a periodic solution with period 2π:

$$x_i = \varphi_i(t), \quad y_i = \psi_i(t),$$

and form the perturbation equations by setting

$$x_i = \varphi_i(t) + \xi_i, \quad y_i = \psi_i(t) + \eta_i.$$

In Chap. 2 we saw that the double integral

$$\iint (dx_1 dy_1 + dx_2 dy_2 + \cdots dx_n dy_n)$$

is an integral invariant, or (which amounts to the same thing) that if ξ_i, η_i, and ξ_i', η_i' are two arbitrary specific solutions of the perturbation equations, then

$$\sum_{i=1}^{n} (\xi_i \eta_i' - \xi_i' \eta_i) = \text{const.}$$

I state that it follows from this that the characteristic exponents are pairwise equal and of opposite sign.

In fact, let ξ_i^0 and η_i^0 be the initial values of ξ_i and η_i for $t = 0$ in one of the perturbation equations; let ξ_i^1 and η_i^1 be the corresponding values of ξ_i and η_i for $t = 2\pi$. It is clear that ξ_i^1 and η_i^1 will be linear functions of the ξ_i^0 and η_i^0 such that the substitution

$$T = \left(\xi_i^0, \eta_i^0; \xi_i^1, \eta_i^1 \right)$$

will be a linear substitution.

Let

$$\begin{vmatrix} a_{11} & a_{12} & \cdots & a_{1,2n} \\ a_{21} & a_{22} & \cdots & a_{2,2n} \\ \vdots & \vdots & \ddots & \vdots \\ a_{2n,1} & a_{2n,2} & \cdots & a_{2n,2n} \end{vmatrix}$$

be the matrix of coefficients for this linear substitution.

Form the equation in λ

$$\begin{vmatrix} a_{11} - \lambda & a_{12} & \cdots & a_{1,2n} \\ a_{21} & a_{22} - \lambda & \cdots & a_{2,2n} \\ \vdots & \vdots & \ddots & \vdots \\ a_{2n,1} & a_{2n,2} & \cdots & a_{2n,2n} - \lambda \end{vmatrix} = 0.$$

The $2n$ roots of this equation will be what are called the $2n$ multipliers of the linear substitution T. But this linear substitution T cannot be arbitrary. It has to leave the bilinear form unaltered:

$$\sum \left(\xi_i \eta_i' - \xi_i' \eta_i \right).$$

To do that, the equation in λ must be reciprocal. If therefore we set

$$\lambda = e^{2\alpha\pi},$$

the quantities α will need to be equal pairwise and of opposite sign.
Which was to be proved.

There will, therefore, in general be n distinct quantities α^2. We will call them the *stability coefficients* of the periodic solution being considered.

If these n coefficients are all real and negative, the periodic solutions will be stable, because the quantities ξ_i and η_i will remain below a given bound.

This word must, however, not be understood to mean stability in the absolute sense. In fact, we have neglected the squares of the ξ and η and nothing proves that the result would not change when these squares are included. But we can state at least that if the ξ and η were very small to begin with, they will remain very small for a very long time. We can express this fact by stating that if the periodic solution does not have *secular* stability, then at least it has *temporary* stability.

This stability can be considered by referring to the values of ξ_i; in fact for the general solution of the perturbation equations, one finds

$$\xi_i = \sum A_k e^{\alpha_k t} S_{ik},$$

where the A_k are constant coefficients and the S_{ik} trigonometric series.

Hence, if α_k^2 is real and negative, one finds

$$e^{\alpha_k t} = \cos t \sqrt{-\alpha_k^2} + i \sin t \sqrt{-\alpha_k^2},$$

such that ξ_i are expressed trigonometrically.

In contrast, if one or more stability coefficients become real and positive or else imaginary, the periodic solution considered no longer enjoys temporary stability.

In fact it can be easily seen that ξ_i is then represented by a series whose general term has the following form:

$$Ae^{ht}\cos(kt+mt+l),$$

where $(h+ik)^2$ is one of the stability coefficients, where m is an integer and l and A are arbitrary constants. The breakdown of stability is thus clearly seen.

If two of the stability coefficients become equal to each other, or if one of them becomes zero, terms of the following form will be in general be found in the series representing ξ_i:

$$Ate^{ht}\cos(kt+mt+l) \quad \text{or} \quad At\cos(mt+l).$$

In summary, ξ_i can in all cases be represented by a series that is always convergent. In this series, time can enter within the sine or cosine, or in the exponential e^{ht}, or also outside trigonometric or exponential functions.

If all the stability coefficients are real, negative, and distinct, time will only appear within the sine and cosine and there will be temporary stability.

If one of the coefficients is positive or imaginary, time will appear in an exponential; if two of the coefficients are equal or if one of them is zero, time will appear outside of any trigonometric or exponential function.

Therefore, if the coefficients are not all real, negative, and distinct, there is in general no temporary stability.

Whenever F does not depend on time t, one of the n stability coefficients is zero, both because the time does not enter explicitly into the differential equations and also because these equations allow an integral

$$F(x_1, x_2, \ldots x_n; y_1, y_2, \ldots y_n) = \text{const.}$$

We therefore find ourselves in the case that we spoke about above and where two of the characteristic exponents are zero. But, as we stated, that does not create a difficulty because using the known integral the order of Eq. (1') can be lowered to $2n-1$. There are then only $2n-1$ characteristic exponents; one of them is zero and the $2n-2$ others, whose squares can still be called stability coefficients, are equal pairwise and of opposite sign.

Let us go back to the determinant that we needed to consider in the previous section.

In that section, we first considered the case where Eq. (1) depends on time t and a parameter of μ, and for $\mu = 0$ allow one and only one periodic solution. We have seen that if the Jacobian determinant

2 Characteristic Exponents

$$\Delta = \frac{\partial(\psi_1, \psi_2, \ldots \psi_n)}{\partial(\beta_1, \beta_2, \ldots \beta_n)} \neq 0,$$

then the equations will still allow a periodic solution for small values of μ.

This determinant can be written as

$$\Delta = \begin{vmatrix} \frac{d\gamma_1}{d\beta_1} - 1 & \frac{d\gamma_1}{d\beta_2} & \cdots & \frac{d\gamma_1}{d\beta_n} \\ \frac{d\gamma_2}{d\beta_1} & \frac{d\gamma_2}{d\beta_2} - 1 & \cdots & \frac{d\gamma_2}{d\beta_n} \\ \vdots & \vdots & \ddots & \vdots \\ \frac{d\gamma_n}{d\beta_1} & \frac{d\gamma_n}{d\beta_2} & \cdots & \frac{d\gamma_n}{d\beta_n} - 1 \end{vmatrix}.$$

Hence, the characteristic exponents α are given by the following equation:

$$\begin{vmatrix} \frac{d\gamma_1}{d\beta_1} - e^{2\alpha\pi} & \frac{d\gamma_1}{d\beta_2} & \cdots & \frac{d\gamma_1}{d\beta_n} \\ \frac{d\gamma_2}{d\beta_1} & \frac{d\gamma_2}{d\beta_2} - e^{2\alpha\pi} & \cdots & \frac{d\gamma_2}{d\beta_n} \\ \vdots & \vdots & \ddots & \vdots \\ \frac{d\gamma_n}{d\beta_1} & \frac{d\gamma_n}{d\beta_2} & \cdots & \frac{d\gamma_n}{d\beta_n} - e^{2\alpha\pi} \end{vmatrix} = 0.$$

Stating that Δ is zero amounts to stating that one of the characteristic exponents is zero such that we can state the first of the theorems proved in the previous section in the following way:

If Eq. (1) that depend on a parameter μ allow a periodic solution for $\mu = 0$ for which none of the characteristic exponents are zero, then it will still allow a periodic solution for small values of μ.

3 Periodic Solutions of the Equations of Dynamics

To be more definite, I will take the equations of dynamics with three degrees of freedom, but what I am going to say is obviously applicable to the general case. I will, therefore, write my equations in the following form:

$$\frac{dx_1}{dt} = \frac{dF}{dy_1}, \quad \frac{dx_2}{dt} = \frac{dF}{dy_2}, \quad \frac{dx_3}{dt} = \frac{dF}{dy_3},$$
$$\frac{dy_1}{dt} = -\frac{dF}{dx_1}, \quad \frac{dy_2}{dt} = -\frac{dF}{dx_2}, \quad \frac{dy_3}{dt} = -\frac{dF}{dx_3}, \quad (1)$$

where F is an arbitrary one-to-one function of x and y, independent of t.

I will next assume that x_1, x_2, and x_3 are linear variables, but that y_1, y_2, and y_3 are angular variables, meaning that F is a periodic function of y_1, y_2, and y_3 with period 2π, such that the state of the system does not change when one or more of the three quantities y increases by a multiple of 2π. (See Chap. 1)

I will further assume that F depends on an arbitrary parameter μ and is expandable in increasing powers of this parameter such that one gets

$$F = F_0 + \mu F_1 + \mu^2 F_2 + \mu^3 F_3 + \cdots$$

I will finally assume that F_0 depends only on the x and is independent of the y such that

$$\frac{dF_0}{dy_1} = \frac{dF_0}{dy_2} = \frac{dF_0}{dy_3} = 0.$$

Nothing is simpler then than integrating Eq. (1) when $\mu = 0$; they are in fact written as

$$\frac{dx_1}{dt} = \frac{dx_2}{dt} = \frac{dx_3}{dt} = 0 \quad \frac{dy_1}{dt} = -\frac{dF_0}{dx_1}, \quad \frac{dy_2}{dt} = -\frac{dF_0}{dx_2}, \quad \frac{dy_3}{dt} = -\frac{dF_0}{dx_3}.$$

These equations show first that x_1, x_2, and x_3 are constants. From this it can be concluded that

$$-\frac{dF_0}{dx_1}, \quad -\frac{dF_0}{dx_2}, \quad -\frac{dF_0}{dx_3},$$

which only depend on x_1, x_2, and x_3 are also constants that we will call for brevity n_1, n_2, and n_3 and which are completely defined when the constant values of x_1, x_2, and x_3 are given. It then follows:

$$y_1 = n_1 t + \varpi_1, \quad y_2 = n_2 t + \varpi_2, \quad y_3 = n_3 t + \varpi_3,$$

where ϖ_1, ϖ_2, and ϖ_3 are new constants of integration.

What is the condition so that the solution thus found will be periodic and of period T. It must be that if t changes to $t + T$, then y_1, y_2, and y_3 increase by a multiple of 2π, which means that

$$n_1 T, \quad n_2 T \quad \text{and} \quad n_3 T$$

are multiples of 2π.

Thus in order for the solution that we just found to be periodic, it is necessary and sufficient that the three numbers n_1, n_2, and n_3 be commensurable with each other.

3 Periodic Solutions of the Equations of Dynamics

As for the period T, it will be the least common multiplier of the three quantities:

$$\frac{2\pi}{n_1}, \quad \frac{2\pi}{n_2} \quad \text{and} \quad \frac{2\pi}{n_3}.$$

We exclude from our searches, at least for now, the case where the three functions dF_0/dx_1, dF_0/dx_2, and dF_0/dx_3 are not independent of each other. If this case is put aside, x_1, x_2, and x_3 can always be chosen in such a way that n_1, n_2, and n_3 have such values as we want, at least in a certain domain. There will, therefore, be infinitely many possible choices for the three constants x_1, x_2, and x_3 which will lead to periodic solutions.

I propose to look for whether there still exist periodic solutions with period T when μ is not equal to 0.

To demonstrate it, I am going to use reasoning analogous to that from Sect. 1.

Suppose that μ stops being zero, and imagine that, in a certain solution, the values of x and y for $t = 0$ are, respectively,

$$x_1 = a_1 + \delta a_1, \quad x_2 = a_2 + \delta a_2, \quad x_3 = a_3 + \delta a_3,$$
$$y_1 = \varpi_1 + \delta\varpi_1, \quad y_2 = \varpi_2 + \delta\varpi_2, \quad y_3 = \varpi_3 + \delta\varpi_3.$$

Assume that, in the same solution, the values of the x and y for $t = T$ are

$$x_1 = a_1 + \delta a_1 + \Delta a_1,$$
$$x_2 = a_2 + \delta a_2 + \Delta a_2,$$
$$x_3 = a_3 + \delta a_3 + \Delta a_3,$$
$$y_1 = \varpi_1 + n_1 T + \delta\varpi_1 + \Delta\varpi_1,$$
$$y_2 = \varpi_2 + n_2 T + \delta\varpi_2 + \Delta\varpi_2,$$
$$y_3 = \varpi_3 + n_3 T + \delta\varpi_3 + \Delta\varpi_3.$$

The condition for this solution to be periodic with period T is that the following holds:

$$\Delta a_1 = \Delta a_2 = \Delta a_3 = \Delta\varpi_1 = \Delta\varpi_2 = \Delta\varpi_3 = 0. \tag{2}$$

The six equations in Eq. (2) are not distinct. In fact, as $F = $ constant is an integral of Eq. (1), and since additionally F is periodic with respect to y, it follows that

$$F(a_i + \delta a_i, \varpi_i + \delta\varpi_i) = F(a_i + \delta a_i + \Delta a_i, \varpi_i + n_i T + \delta\varpi_i + \Delta\varpi_i)$$
$$= F(a_i + \delta a_i + \Delta a_i, \varpi_i + \delta\varpi_i + \Delta\varpi_i).$$

It will, therefore, be sufficient for us to satisfy five of Eq. (2). I will additionally suppose

$$\varpi_1 = \delta\varpi_1 = 0,$$

which amounts to taking the epoch where y_1 is zero for the origin of time. It is easy to see that the Δa_i and the $\Delta\varpi_i$ are mapping functions of μ, the δa_i and the $\delta\varpi_i$ becoming zero when all these variables become zero.

It is then a matter of proving that δa_1, δa_2, δa_3, $\delta\varpi_2$, and $\delta\varpi_3$ can be drawn as functions of μ from these last five of Eq. (2).

Remark that when μ is zero, it holds that

$$\Delta a_1 = \Delta a_2 = \Delta a_3 = 0.$$

Consequently Δa_1, Δa_2, and Δa_3, expanded in powers of μ from δa_i and $\delta\varpi_i$, contain μ as a factor. We will eliminate this factor μ, and we will consequently write the five of Eq. (2) that we need to solve in the form:

$$\frac{\Delta a_2}{\mu} = \frac{\Delta a_3}{\mu} = \Delta\varpi_1 = \Delta\varpi_2 = \Delta\varpi_3 = 0. \tag{3}$$

It remains for us to determine ϖ_2 and ϖ_3 such that these equations are satisfied for

$$\mu = \delta\varpi_2 = \delta\varpi_3 = \delta a_1 = \delta a_2 = \delta a_3 = 0. \tag{4}$$

Let us look at what the first terms of Eq. (3) become when setting $\mu = 0$ in them. It follows that

$$n_1 T + \Delta\varpi_1 = + \int_0^T \frac{dy_1}{dt} dt = -\int_0^T \frac{dF}{dx_1} dt = -\int_0^T \frac{dF_0}{d(a_1 + \delta a_1)} dt,$$

hence,

$$\Delta\varpi_1 = -T\left(\frac{dF_0}{dx_1} + n_1\right)$$

and similarly,

$$\Delta\varpi_2 = -T\left(\frac{dF_0}{dx_2} + n_2\right),$$

$$\Delta\varpi_3 = -T\left(\frac{dF_0}{dx_3} + n_3\right).$$

3 Periodic Solutions of the Equations of Dynamics

It is important to observe that in F_0, x_1, x_2, and x_3 need to be replaced by $a_1 + \delta a_1$, $a_2 + \delta a_2$, and $a_3 + \delta a_3$; in fact for $\mu = 0$, F reduces to F_0 and x_1, x_2, and x_3 to constants which remain continually equal to their initial values $a_1 + \delta a_1$, $a_2 + \delta a_2$, and $a_3 + \delta a_3$.

It also follows that

$$\frac{\Delta a_2}{\mu} = \frac{1}{\mu}\int_0^T \frac{dx_2}{dt}dt = \frac{1}{\mu}\int_0^T \frac{dF}{dy_2}dt$$

or because F_0 does not depend on y_2:

$$\frac{\Delta a_2}{\mu} = \int_0^T \frac{d}{dy_2}\left(\frac{F-F_0}{\mu}\right)dt$$

or for $\mu = 0$

$$\frac{\Delta a_2}{\mu} = \int_0^T \frac{dF_1}{dy_2}dt.$$

Let us assume that μ, the $\delta\varpi$, and the δa are zero at the same time; it would then be necessary to make in F_1

$$x_1 = a_1, \quad x_2 = a_2, \quad x_3 = a_3,$$
$$y_1 = n_1 t, \quad y_2 = n_2 t + \varpi_2, \quad y_3 = n_3 t + \varpi_3.$$

F_1 will then become a periodic function of t with period T, and a periodic function of ϖ_2 and ϖ_3 with period 2π.

Let ψ be the average value of F_1 considered as a periodic function of t. It follows that

$$\frac{\Delta a_2}{\mu} = \int_0^T \frac{dF_1}{d\varpi_2}dt = T\frac{d\psi}{d\varpi_2}$$

and similarly

$$\frac{\Delta a_3}{\mu} = T\frac{d\psi}{d\varpi_3}.$$

We therefore have to choose ϖ_2 and ϖ_3 so as to satisfy the equations

$$\frac{d\psi}{d\varpi_2} = \frac{d\psi}{d\varpi_3} = 0. \tag{5}$$

This is always possible; in fact the function ψ is periodic in ϖ_2 and ϖ_3 and it is finite; therefore it has at least one maximum and one minimum for which these two derivatives must become zero. Once ϖ_2 and ϖ_3 have been chosen that way, it will be seen that Eq. (3) is satisfied when the following are set at the same time:

$$\mu = \delta\varpi_2 = \delta\varpi_3 = \delta a_1 = \delta a_2 = \delta a_3 = 0.$$

We can, therefore, draw from Eq. (3) the five unknowns δa_i and $\delta \varpi_i$ in the form of mapping functions of μ, which become zero with μ. There will only be an exception if the Jacobian determinant

$$\frac{\partial\left(\frac{\Delta a_2}{\mu}, \frac{\Delta a_3}{\mu}, \Delta\varpi_1, \Delta\varpi_2, \Delta\varpi_3\right)}{\partial(\delta a_1, \delta a_2, \delta a_3, \delta\varpi_2, \delta\varpi_3)}$$

were to be zero. But for $\mu = 0$, $\Delta\varpi_1$, $\Delta\varpi_2$, and $\Delta\varpi_3$ are independent of ϖ_2 and ϖ_3 such that this Jacobian determinant is the product of two others:

$$\frac{\partial\left(\frac{\Delta a_2}{\mu}, \frac{\Delta a_3}{\mu}\right)}{\partial(\delta\varpi_2, \delta\varpi_3)} \quad \text{and} \quad \frac{\partial(\Delta\varpi_1, \Delta\varpi_2, \Delta\varpi_3)}{\partial(\delta a_1, \delta a_2, \delta a_3)}.$$

If the factors T^2 and $-T^3$ are dropped, the first of these determinants is equal to the Hessian determinant of ψ relative to ϖ_2 and ϖ_3 and the second to the Hessian determinant of F_0 relative to x_1^0, x_2^0 and x_3^0.

If therefore neither of these Hessian determinants is zero, it will be possible to satisfy the five of Eq. (3) and consequently for sufficiently small values of μ, there will exist a periodic solution with period T.
Which was to be proved.

We are now going to try to determine, no longer just the periodic solutions with period T, but the solutions with period slightly different from T. For a starting point, we took the three numbers n_1, n_2, n_3; we could just as well have chosen three other numbers n_1', n_2', n_3', provided that they were commensurable with each other, and we would have arrived at a periodic solution whose period T would have been the least common multiple of $2\pi/n_1', 2\pi/n_2', 2\pi/n_3'$.

If, in particular, we take

$$n_1' = n_1(1+\varepsilon), \quad n_2' = n_2(1+\varepsilon), \quad n_3' = n_3(1+\varepsilon);$$

the three numbers n_1', n_2', n_3' will be commensurable with each other because they are proportional to the three numbers n_1, n_2, n_3.

3 Periodic Solutions of the Equations of Dynamics

This will, therefore, lead us to a periodic solution with period:

$$T' = \frac{1+\varepsilon}{T}$$

such that we will have

$$x_i = \varphi_i(t, \mu, \varepsilon), \quad y_i = \varphi'_i(t, \mu, \varepsilon), \tag{6}$$

where the φ_i and the φ'_i are functions which are expandable in powers of μ and ε, and periodic in t, but such that the period depends on ε.

If in F we replace the x_i and the y_i by their values (4), F must become a time-independent constant [because $F =$ constant is one of the integrals of Eq. (1)]. But this constant, which is the total energy, will depend on μ and ε and could be expanded in increasing powers of these variables.

If the total energy constant B is a given of the problem, the equation

$$F(\mu, \varepsilon) = B$$

can be regarded as a relation which connects ε to μ. If therefore we arbitrarily choose B,, there will always exist a periodic solution whatever the value chosen for this constant, but the period will depend on ε and consequently on μ.

A more specific case than the one that we just dealt with in detail is the one where there are only two degrees of freedom. F then only depends on four variables x_1, y_1, x_2, y_2 and the function ψ now only depends on a single variable ϖ_2. The relations (5) then reduce to

$$\frac{d\psi}{d\varpi_2} = 0 \tag{7}$$

and the Hessian determinant of ψ is reduced to $d^2\psi/d\varpi_2^2$, from which this conclusion follows:

For each of the single roots of Eq. (7), there corresponds a periodic solution of Eq. (1), which exists for all sufficiently small values of μ.

I could even add that it is even the same for each of the odd-order roots as we have seen in Sect. 1 and that this equation always has similar roots because the function ψ has at least one maximum which can only correspond to the odd roots of Eq. (7).

Let us go back to the case where there are three degrees of freedom and where the period is constant and equal to T.

I state that $x_1, x_2, x_3, y_1, y_2, y_3$ are expandable in increasing powers of μ. In fact, because of Theorem III from Sect. 2 in Chap. 1, the x and y are expandable in powers of μ and of $\delta a_1, \delta a_2, \delta a_3, \delta \varpi_2$, and $\delta \varpi_3$. But let us imagine that the δa and $\delta \varpi$ have been determined such that the solution is periodic with period T. The δa

and $\delta\varpi$ will then be drawn from Eq. (3) in the form of series ordered in powers of μ, such that the x and the y will finally be ordered in powers of μ.

Since the solution needs to be periodic with period T whatever μ, the coefficients of the various powers of μ will be periodic functions of t.

Additionally, let us remark that it can always be assumed that the origin of time has been chosen such that y_1 is zero when t is zero, and that this takes place for any μ. Then for $t = 0$, it will hold

$$0 = y_1^0 = y_1^1 = y_1^2 = \cdots.$$

The existence and convergence of these series having thus been established, I am going to determine the coefficients.

To do that, I am going to seek to satisfy Eq. (1) by making[1]

$$\begin{aligned}
x_1 &= x_1^0 + \mu x_1^1 + \mu^2 x_1^2 + \cdots, \\
x_2 &= x_2^0 + \mu x_2^1 + \mu^2 x_2^2 + \cdots, \\
x_3 &= x_3^0 + \mu x_3^1 + \mu^2 x_3^2 + \cdots, \\
y_1 &= y_1^0 + \mu y_1^1 + \mu^2 y_1^2 + \cdots, \\
y_2 &= y_2^0 + \mu y_2^1 + \mu^2 y_2^2 + \cdots, \\
y_3 &= y_3^0 + \mu y_3^1 + \mu^2 y_3^2 + \cdots.
\end{aligned} \qquad (8)$$

In these formulas, x_1^0, x_2^0, x_3^0 designate the constant values that I had been led above to assign to x_1, x_2, and x_3 when I assumed $\mu = 0$ and which are such that

$$\frac{d}{dx_1^0}F_0(x_1^0,x_2^0,x_3^0) = -n_1, \quad \frac{d}{dx_2^0}F_0(x_1^0,x_2^0,x_3^0) = -n_2, \quad \frac{d}{dx_3^0}F_0(x_1^0,x_2^0,x_3^0) = -n_3.$$

Additionally, it holds that

$$y_i^0 = n_i t + \varpi_i.$$

Finally the x_i^1, the y_i^1, the x_i^2, the y_i^2, etc., are functions of time that it will be a matter of determining and which will need to be periodic with period T.

In F, in place of the x and y, let us substitute their values (8), and then let us expand F according to increasing powers of μ such that we have

$$F = \Phi_0 + \mu\Phi_1 + \mu^2\Phi_2 + \cdots.$$

---

[1] The digits placed to the upper right of the letters x and y in Eq. (8) are indices and not exponents.

3 Periodic Solutions of the Equations of Dynamics

It is clear that

$$\Phi_0 = F_0\left(x_1^0, x_2^0, x_3^0\right)$$

only depends on the x_i^0; that

$$\Phi_1 = F_1\left(x_1^0, x_2^0, x_3^0, y_1^0, y_2^0, y_3^0\right) + x_1^1 \frac{dF_0}{dx_1^0} + x_2^1 \frac{dF_0}{dx_2^0} + x_3^1 \frac{dF_0}{dx_3^0} \quad (9)$$

only depends on the x_i^0, the y_i^0 and the x_i^1; that Φ_2 only depends on the x_i^0, the y_i^0, the x_i^1, the y_i^1, and the x_i^2; etc.

More generally, I can write

$$\Phi_k = \Theta_k + x_1^k \frac{dF_0}{dx_1^0} + x_2^k \frac{dF_0}{dx_2^0} + x_3^k \frac{dF_0}{dx_3^0} = \Theta_k - n_1 x_1^k - n_2 x_2^k - n_3 x_3^k,$$

where Θ_k depends only on

$$x_i^0, \quad x_i^1, \quad \ldots \quad \text{and} \quad x_i^{k-1},$$
$$y_i^0, \quad y_i^1, \quad \ldots \quad \text{and} \quad y_i^{k-1}.$$

I can add that relative to the function Θ_k is a periodic function of y_1^0, y_2^0, y_3^0 with period 2π. Equation (9) shows that $\Theta_1 = F_1$.

With that said, the differential equations can be written by equating the powers of μ of the same order:

$$\frac{dx_1^0}{dt} = \frac{dx_2^0}{dt} = \frac{dx_3^0}{dt} = 0, \quad \frac{dy_1^0}{dt} = n_1, \quad \frac{dy_2^0}{dt} = n_2, \quad \frac{dy_3^0}{dt} = n_3.$$

We next find

$$\frac{dx_1^1}{dt} = \frac{dF_1}{dy_1^0}, \quad \frac{dx_2^1}{dt} = \frac{dF_1}{dy_2^0}, \quad \frac{dx_3^1}{dt} = \frac{dF_1}{dy_3^0} \quad (10)$$

and

$$\frac{dy_1^1}{dt} = -\frac{d\Phi_1}{dx_1^0}, \quad \frac{dy_2^1}{dt} = -\frac{d\Phi_1}{dx_2^0}, \quad \frac{dy_3^1}{dt} = -\frac{d\Phi_1}{dx_3^0}, \quad (11)$$

and more generally,

$$\frac{dx_i^k}{dt} = \frac{d\Phi_k}{dy_i^0} \quad (10')$$

and

$$\frac{dy_i^k}{dt} = -\frac{d\Phi_k}{dx_i^0} = -\frac{d\Theta_k}{dx_i^0} - x_1^k \frac{d^2 F_0}{dx_1^0 dx_i^0} - x_2^k \frac{d^2 F_0}{dx_2^0 dx_i^0} - x_3^k \frac{d^2 F_0}{dx_3^0 dx_i^0}. \quad (11')$$

Let us first integrate Eq. (10). In F_1, we will replace y_1^0, y_2^0, y_3^0 by their values:

$$n_1 t + \varpi_1, \quad n_2 t + \varpi_2, \quad n_3 t + \varpi_3.$$

Because y_1^0 must become zero when t does, ϖ_1 will be zero. Then the right-hand sides of Eq. (10) are periodic functions of t with period T; These right-hand sides can, therefore, be expanded in series proceeding according to the sines and cosines of multiples of $2\pi t/T$. For the values of x_1^1, x_2^1, and x_3^1 drawn from Eq. (10) to be periodic functions of t, it is necessary and sufficient that these series not contain any constant terms.

I can in fact write

$$F_1 = \sum A \sin(m_1 y_1^0 + m_2 y_2^0 + m_3 y_3^0 + h),$$

where m_1, m_2, m_3 are positive or negative integers and where A and h are functions of x_1^0, x_2^0, x_3^0. For brevity, I will write

$$F_1 = \sum A \sin \omega$$

by setting

$$\omega = m_1 y_1^0 + m_2 y_2^0 + m_3 y_3^0 + h.$$

I will then find

$$\frac{dF_1}{dy_1^0} = \sum A m_1 \cos \omega, \quad \frac{dF_1}{dy_2^0} = \sum A m_2 \cos \omega, \quad \frac{dF_1}{dy_3^0} = \sum A m_3 \cos \omega$$

and

$$\omega = t(m_1 n_1 + m_2 n_2 + m_3 n_3) + h + m_2 \varpi_2 + m_3 \varpi_3.$$

Among the terms from the series, I will distinguish those for which

$$m_1 n_1 + m_2 n_2 + m_3 n_3 = 0$$

and which are independent of t. These terms exist because we have assumed that the three numbers n_1, n_2, and n_3 are commensurable between them.

3 Periodic Solutions of the Equations of Dynamics

I will then set

$$\psi = \sum_S A \sin \omega \quad (\text{with} \quad m_1 n_1 + m_2 n_2 + m_3 n_3 = 0, \quad \omega = h + m_2 \varpi_2 + m_3 \varpi_3),$$

where the summation represented by the sign $\sum_S$ extends to all terms of F_1 for which the coefficient of t is zero. We will then have

$$\frac{d\psi}{d\varpi_2} = \sum_S A m_2 \cos \omega, \quad \frac{d\psi}{d\varpi_3} = \sum_S A m_3 \cos \omega.$$

If therefore one has

$$\frac{d\psi}{d\varpi_2} = \frac{d\psi}{d\varpi_3} = 0, \tag{12}$$

it will follow

$$\sum_S A m_1 \cos \omega = 0, \quad \sum_S A m_2 \cos \omega = 0, \quad \sum_S A m_3 \cos \omega = 0. \tag{13}$$

The first of Eq. (11) is in fact a consequence of the two others, because as a result of the relationship $m_1 n_1 + m_2 n_2 + m_3 n_3 = 0$, one has identically

$$n_1 \sum_S A m_1 \cos \omega + n_2 \sum_S A m_2 \cos \omega + n_3 \sum_S A m_3 \cos \omega = 0.$$

If therefore the relationships (12) are satisfied, the series $\sum A m_i \cos \omega$ will not contain a constant term, and Eq. (10) will give us

$$x_1^1 = \sum \frac{A m_1 \sin \omega}{m_1 n_1 + m_2 n_2 + m_3 n_3} + C_1^1, \quad x_2^1 = \sum \frac{A m_2 \sin \omega}{m_1 n_1 + m_2 n_2 + m_3 n_3} + C_2^1,$$

$$x_3^1 = \sum \frac{A m_3 \sin \omega}{m_1 n_1 + m_2 n_2 + m_3 n_3} + C_3^1,$$

where C_1^1, C_2^1, and C_3^1 are three new constants of integration.

It remains for me to prove that the constants ϖ_2 and ϖ_3 can be chosen so as to satisfy the relationships (10). The function ψ is a periodic function of ϖ_2 and ϖ_3 which does not change when one of these two variables increases by 2π. Additionally, it is finite, so it will have at least one maximum and one minimum. There are, therefore, at least two ways to choose ϖ_2 and ϖ_3 so as to satisfy the relationships (12).

I could even add that there are at least four of them without, however, being able to affirm that it is still so when there are more than three degrees of freedom.

I am now going to try to determine the three functions y_i^1 and the three constants C_i^1 using Eq. (11).

We can view the x_i^0 and y_i^0 as known; the x_i^1 are also known up to the constants C_i^1. I can therefore write Eq. (11) in the following form:

$$\frac{dy_i^1}{dt} = H_i - C_1^1 \frac{d^2 F_0}{dx_1^0 dx_i^0} - C_2^1 \frac{d^2 F_0}{dx_2^0 dx_i^0} - C_3^1 \frac{d^2 F_0}{dx_3^0 dx_i^0}, \tag{14}$$

where the H_i represents completely known functions expanded in series of sines and cosines of multiples of $2\pi t/T$. The coefficients C_1^1, C_2^1, and C_3^1 are constants that can be regarded as known.

In order for the value of y_i^1 drawn from this equation to be a periodic function of t, it is necessary and sufficient that the constant term in the right-hand side be zero. If therefore H_i^0 designates the constant term from the trigonometric series H_i, I would need to have

$$C_1^1 \frac{d^2 F_0}{dx_1^0 dx_i^0} + C_2^1 \frac{d^2 F_0}{dx_2^0 dx_i^0} + C_3^1 \frac{d^2 F_0}{dx_3^0 dx_i^0} = H_i^0. \tag{15}$$

The three linear equations in Eq. (15) determine the three constants C_1^1, C_2^1, and C_3^1.

There would only be an exception if the determinant of these three equations were zero; meaning if the Hessian determinant of F_0 relative to the x_1^0, x_2^0, and x_3^0 was zero; we will exclude this case.

Equation (14) thus will give me

$$y_1^1 = \eta_1^1 + k_1^1, \quad y_2^1 = \eta_2^1 + k_2^1, \quad y_3^1 = \eta_3^1 + k_3^1,$$

where the η_i^1 are periodic functions of t, completely known and approaching zero when t does, and where the k_i^1 are three new constants of integration.

Let us now come to Eq. (10′) in y making $k = 2$ and $i = 1, 2, 3$ and let us try to determine with the help of these three resulting equations, the three functions x_i^2, and the three constants k_i^1.

It is easy to see that we have

$$\Theta_2 = \Omega_2 + y_1^1 \frac{dF_1}{dy_1^0} + y_2^1 \frac{dF_1}{dy_2^0} + y_3^1 \frac{dF_1}{dy_3^0},$$

where Ω_2 depends only on the x_i^0, the y_i^0, and the x_i^1, and where we have, as above:

$$\frac{dF_1}{dy_i^0} = \sum Am_i \cos \omega.$$

3 Periodic Solutions of the Equations of Dynamics

Equations (10′) are then written as

$$\frac{dx_i^2}{dt} = \frac{d\Omega_2}{dy_i^0} + \sum_k y_k^1 \frac{d^2 F_1}{dy_k^0 dy_i^0}.$$

or

$$\frac{dx_i^2}{dt} = H_i' - k_1^1 \sum Am_1 m_i \sin \omega - k_2^1 \sum Am_2 m_i \sin \omega - k_3^1 \sum Am_3 m_i \sin \omega, \quad (16)$$

where H_i' is a periodic function of t, which can be regarded as fully known. So x_i^2 can be drawn from this equation in the form of a periodic function, it is necessary and sufficient that the right-hand sides of Eq. (16), expanded in trigonometric series, have no constant terms. We must, therefore, have quantities k_i^1 so as to cancel these constant terms. We would thus be led to three linear equations among the three quantities k_i^1; but since the determinant of these three equations is zero, there is a small difficulty and I will be forced to go into some details.

As y_1^1 approaches 0 with t, it must be that

$$k_1^1 = 0;$$

we will then have only two unknowns k_2^1 and k_3^1 and three equations to be satisfied, but these three equations are not distinct as we are going to see.

In fact, call E_i the constant term from H_i'; these three equations will be written as

$$E_1 = k_2^1 \sum_S Am_2 m_1 \sin \omega + k_3^1 \sum_S Am_3 m_1 \sin \omega,$$
$$E_2 = k_2^1 \sum_S Am_2 m_2 \sin \omega + k_3^1 \sum_S Am_3 m_2 \sin \omega, \quad (17)$$
$$E_3 = k_2^1 \sum_S Am_2 m_3 \sin \omega + k_3^1 \sum_S Am_3 m_3 \sin \omega,$$

while keeping the meaning given above to the summation sign $\sum_S$. I will first only consider the last two Eq. (17) that I will write

$$-E_2 = k_2^1 \frac{d^2 \psi}{d\varpi_2 d\varpi_2} + k_3^1 \frac{d^2 \psi}{d\varpi_3 d\varpi_2},$$
$$-E_3 = k_2^1 \frac{d^2 \psi}{d\varpi_2 d\varpi_3} + k_3^1 \frac{d^2 \psi}{d\varpi_3 d\varpi_3}.$$

From these equations, k_2^1 and k_3^1 can be drawn, unless the Hessian of ψ relative to ϖ_2 and ϖ_3 is zero. If the resulting values are given to k_i^1, the last two Eq. (16) will give us x_2^2 and x_3^2 in the following form:

$$x_2^2 = \xi_2^2 + C_2^2, \quad x_3^2 = \xi_3^2 + C_3^2,$$

where the ξ_i^2 are fully known periodic functions of t and the C_i^2 are new constants of integration.

To find x_1^2 we can, instead of using the first of Eq. (16), make use of the following considerations:

Equation (1) has an integral:

$$F = B,$$

where B is a constant of integration that I will assume to be expanded in powers of μ by writing

$$B = B_0 + \mu B_1 + \mu^2 B_2 + \cdots,$$

such that one has

$$\Phi_0 = B_0, \quad \Phi_1 = B_1, \quad \Phi_2 = B_2, \quad \ldots,$$

depends on the x_i^0, the y_i^0, the x_i^1, the y_i^1, x_2^2, and x_3^2 which are known functions of t and of x_1^2 which we have not yet calculated. From this equation, we can therefore draw x_1^2 in the following form:

$$x_1^2 = \xi_1^2 + C_1^2.$$

ξ_1^2 will be a completely determined periodic function of t and C_1^2 is a constant which depends on B_2, C_2^2, and C_2^3.

From this we can conclude that the first of Eq. (17) must be satisfied and consequently that these three equations of Eq. (17) are not distinct.

Now take Eq. (11′) and in them set $k = 2$; we will get three equations that will allow us to determine the constants C_1^1, C_1^2, and C_1^3 and from which furthermore the y_i^2 will be drawn in the following form:

$$y_1^2 = \eta_1^2 + k_1^2, \quad y_2^2 = \eta_2^2 + k_2^2, \quad y_3^2 = \eta_3^2 + k_3^2,$$

where the η are completely known periodic functions of t and the k_i^2 are three new constants of integration.

Next picking back up with Eq. (10′) by setting $k = 3$ in them; if we assume that $k_1^2 = 0$, we will be able draw from the three resulting equations, first the two constants k_2^2 and k_3^2, and then the x_i^3 in the following form:

$$x_i^3 = \xi_i^3 + C_i^3,$$

3 Periodic Solutions of the Equations of Dynamics

where the ξ are known periodic functions of t and where the C_i^3 are three new constants of integration,

And so on.

And that is a procedure for finding series ordered by powers of μ, periodic with period T with respect to time and satisfying Eq. (1). *This procedure would fail only if the Hessian of F_0 relative to the x_i^0 were zero or if the Hessian of ψ relative to ϖ_2 and ϖ_3 were zero.*

What we just stated applies in particular to an equation which is sometimes encountered in celestial mechanics and with which several mathematicians are already involved. The following is the equation:

$$\frac{d^2\rho}{dt^2} + n^2\rho + m\rho^3 = \mu R(\rho,t), \tag{18}$$

where n and m are constants, μ is a very small parameter, and R is a function of ρ and t, expanded in increasing powers of ρ and periodic with respect to t.

For us to properly incorporate this, Eq. (18) must first be brought into the canonical form of the equations of dynamics. This can be done by setting

$$\xi = t, \quad \frac{d\rho}{dt} = \sigma, \quad F = \frac{\sigma^2}{2} + \frac{n^2\rho^2}{2} + \frac{m\rho^4}{4} - \mu \int R(\rho,\xi)d\rho + \eta,$$

where ξ and η are two new auxiliary variables and where the integral $\int R(\rho,\xi)d\rho$ is calculated by regarding the ξ as a constant. It is then found

$$\frac{d\rho}{dt} = \frac{dF}{d\sigma}, \quad \frac{d\sigma}{dt} = \frac{dF}{d\rho}, \quad \frac{d\xi}{dt} = \frac{dF}{d\eta}, \tag{19}$$

to which we will be able to add (since so far η has remained completely arbitrary) the following equation:

$$\frac{d\eta}{dt} = -\frac{dF}{d\xi}, \tag{19}$$

which completes a canonical system.

When $\mu = 0$ the general integral of Eq. (18) is written as

$$\rho = h\,\mathrm{sn}(gt+\varpi), \quad \sigma = hg\,\mathrm{cn}(gt+\varpi)\mathrm{dn}(gt+\varpi), \tag{20}$$

where g and ϖ are two constants of integration and where h, and also the modulus of the amplitude sine, are both easily determined functions of g.

We are going to change variables; instead of ξ, η, ρ and σ, we are going to take four variables x_1, y_1, x_2, y_2 defined as follows. We will have first

$$x_2 = \eta, \quad y_2 = \xi.$$

From Eq. (20) which give ρ and σ as functions of g and $gt + \varpi$ for $\mu = 0$, we can get g and $gt + \varpi$ as functions of ρ and σ. It follows

$$g = \chi_1(\rho, \sigma), \quad gt + \varpi = \chi_2(\rho, \sigma).$$

We will then take for x_1 some function of $\chi_1(\rho, \sigma)$ and for y_1

$$y_1 = \frac{k}{2\pi} \chi_2(\rho, \sigma),$$

where k designates the real period of $sn(x)$.

If x_1 has then been suitably chosen as a function of χ_1, the equations will retain their canonical form

$$\frac{dy_1}{dt} = \frac{dF}{dx_1}, \quad \frac{dy_2}{dt} = \frac{dF}{dx_2}, \quad \frac{dx_1}{dt} = -\frac{dF}{dy_1}, \quad \frac{dx_2}{dt} = -\frac{dF}{dy_2}.$$

It is furthermore clear that for $\mu = 0$, F depends only on x_1 and x_2 and not on y_1 and y_2.

We then find ourselves back in the conditions stated at the beginning of this section.

Equation (18) was especially studied by mathematicians in the case where $m = 0$; at first glance it appears that it is then much simpler. This is only an illusion; in fact, if one assumes that $m = 0$, one enters the case where the Hessian of F_0 is zero and what we stated in the previous section is no longer applicable without modification.

This does not mean that the specifics presented by Eq. (18) in the general case are no longer true for $m = 0$, every time at least that μ is not zero. The only difference is that they can no longer be proved by an expansion in powers of μ. The apparent simplification thus made to Eq. (18) has only increased the difficulties. It is true that when $m = 0$ one is led to much simpler series than in the general case, but these series do not converge as we will see in the following.

The method presented in this section also applies to a specific case of the three-body problem.

Assume a zero mass C attracted by two moving masses A and B, one equal to $1 - \mu$ and the other to μ, and describing with a continuous motion two concentric circumferences around the mutual center of gravity which is assumed to be fixed. Additionally, imagine that the mass C moves in the plane of these two circumferences.

Further on, we will see that in this case the equations of motion can be put in the following form:

3 Periodic Solutions of the Equations of Dynamics

$$\frac{dx_1}{dt} = \frac{dF}{dy_1}, \quad \frac{dx_2}{dt} = \frac{dF}{dy_2},$$
$$\frac{dy_1}{dt} = -\frac{dF}{dx_1}, \quad \frac{dy_2}{dt} = -\frac{dF}{dx_2}. \tag{1}$$

Here, x_1 designates the areal speed of the point C, x_2 designates the square root of the major axis of the orbit of C, y_1 designates the difference between the longitude of perihelion of C and the longitude of B, and y_2 designates the mean anomaly.

Additionally, F is expandable in powers of μ and it holds that

$$F_0 = x_1 + \frac{1}{2x_2^2}.$$

It is then easy to see that the Hessian of F_o relative to x_1 and x_2 is zero. It therefore appears that the methods of this section do not work. It is nothing of the sort and a very simple trick can be used to overcome the difficulty.

Equation (1) has an integral

$$F = C.$$

Let us consider the constant C as a given of the problem.

If then $\varphi(F)$ is an arbitrary function of F and if $\varphi'(F)$ is its derivative, then it will hold

$$\varphi'(F) = \varphi'(C)$$

and Eq. (1) can be written as

$$\frac{dx_i}{dt} = \frac{\varphi'(F)}{\varphi'(C)} \frac{dF}{dy_i}, \quad \frac{dy_i}{dt} = -\frac{\varphi'(F)}{\varphi'(C)} \frac{dF}{dx_i}$$

or

$$\frac{dx_i}{dt} = \frac{d}{dy_i}\left[\frac{\varphi(F)}{\varphi'(C)}\right], \quad \frac{dy_i}{dt} = -\frac{d}{dx_i}\left[\frac{\varphi(F)}{\varphi'(C)}\right]. \tag{1'}$$

In general, the Hessian of $\varphi(F_0)/\varphi'(C)$ will not be zero. This is what happens in particular when

$$\varphi(F_0) = F_0^2 = x_1^2 + \frac{x_1}{x_2^2} + \frac{1}{4x_2^4}.$$

The solutions of Eq. (1) which correspond to the specific value C of the integral F also belong to Eq. (1').

Now let us consider a solution of Eq. (1) which is such that the integral F is equal to a constant C_1 different from C.

I state that this solution will still belong to Eq. (1') provided that in it t is changed to

$$t \frac{\varphi'(C_1)}{\varphi'(C)}.$$

One has in fact

$$\frac{dx_i}{dt} = \frac{dF}{dy_i}, \quad \frac{dy_i}{dt} = -\frac{dF}{dx_i};$$

if t is changed into $t\varphi'(C_1)/\varphi'(C)$, it will follow

$$\frac{dx_i}{dt} = \frac{\varphi'(C_1)}{\varphi'(C)} \frac{dF}{dy_i}, \quad \frac{dy_i}{dt} = -\frac{\varphi'(C_1)}{\varphi'(C)} \frac{dF}{dx_i}$$

or because $F = C_1$

$$\frac{dx_i}{dt} = \frac{\varphi'(F)}{\varphi'(C)} \frac{dF}{dy_i}, \quad \frac{dy_i}{dt} = -\frac{\varphi'(F)}{\varphi'(C)} \frac{dF}{dx_i}.$$

Which was to be proved.

From the solutions for Eq. (1), it is easy to deduce those for Eq. (1') and vice versa.

The methods from this section are therefore, because of this trick, applicable to this specific case of the three-body problem.

They will not be so easily applicable to the general case. In the general case in fact, not only is the Hessian of F_0 zero, but that of $\varphi(F_0)$ is also zero, whatever the function φ.

From there, there are some difficulties that I will not discuss here; I will return to them later and for the moment I will limit myself to referring the reader to work that I published in the *Bulletin astronomique*, volume 1, page 65.

4 Calculation of the Characteristic Exponents

Return to Eq. (1) from the previous section.

$$\frac{dx_i}{dt} = \frac{dF}{dy_i}, \quad \frac{dy_i}{dt} = -\frac{dF}{dx_i}. \quad (i = 1, 2, 3) \tag{1}$$

4 Calculation of the Characteristic Exponents

Let us assume that a periodic solution of these equations has been found:

$$x_i = \varphi_i(t), \quad y_i = \psi_i(t)$$

and let us propose to determine the characteristic exponents of this solution.

To do that, we will set

$$x_i = \varphi_i(t) + \xi_i, \quad y_i = \psi_i(t) + \eta_i,$$

and then we will form the perturbation equations of Eq. (1) that we will write as

$$\frac{d\xi_i}{dt} = \sum_k \frac{d^2 F}{dy_i dx_k} \xi_k + \sum_k \frac{d^2 F}{dy_i dy_k} \eta_k,$$
$$\frac{d\eta_i}{dt} = -\sum_k \frac{d^2 F}{dx_i dx_k} \xi_k - \sum_k \frac{d^2 F}{dx_i dy_k} \eta_k, \quad (i, k = 1, 2, 3) \tag{2}$$

and then we will seek to integrate these equations by setting

$$\xi_i = e^{at} S_i, \quad \eta_i = e^{at} T_i, \tag{3}$$

where S_i and T_i are periodic functions of t. We know that in general there exist six specific solutions of this form [since the linear equations in Eq. (2) are of sixth order]. But it is important to note that in the specific case that concerns us, there are no more than four specific solutions which retain this form, because two of the characteristic exponents are zero, and consequently there are two specific solutions of degenerate form.

With that said, let us first assume $\mu = 0$, then F reduces to F_0 as we have seen in the previous section and now only depends on x_1^0, x_2^0, and x_3^0.

Then Eq. (2) reduces to

$$\frac{d\xi_i}{dt} = 0, \quad \frac{d\eta_i}{dt} = -\sum_k \frac{d^2 F_0}{dx_i^0 dx_k^0} \xi_k. \tag{2'}$$

The coefficients of ξ_k in the second Eq. (2') are constants.

For solutions of Eq. (2'), we will take

$$\xi_1 = \xi_2 = \xi_3 = 0, \quad \eta_1 = \eta_1^0, \quad \eta_2 = \eta_2^0, \quad \eta_3 = \eta_3^0,$$

where η_1^0, η_2^0, and η_3^0 are three constants of integration.

This solution is not the most general because it only contains three arbitrary constants, but it is the most general among those that can be written in the form (3). We thus see that for $\mu = 0$, the six characteristic exponents are zero.

Now, no longer assume that μ is zero. We are now going to try to expand α, S_i, and T_i, not according to increasing powers of μ, but according to increasing powers of $\sqrt{\mu}$, by writing

$$\alpha = \alpha_1\sqrt{\mu} + \alpha_2\mu + \alpha_3\mu\sqrt{\mu} + \cdots,$$
$$S_i = S_i^0 + S_i^1\sqrt{\mu} + S_i^2\mu + S_i^3\mu\sqrt{\mu} + \cdots,$$
$$T_i = T_i^0 + T_i^1\sqrt{\mu} + T_i^2\mu + T_i^3\mu\sqrt{\mu} + \cdots.$$

I first propose to establish that this expansion is possible.

We start by showing that the characteristic exponents α are expandable in increasing powers of $\sqrt{\mu}$.

According to what we saw in Sect. 1, the characteristic exponents will be given by the following equation, using the notation from Sects. 1 and 2:

$$\begin{vmatrix} \dfrac{d\gamma_1}{d\beta_1} - e^{\alpha T} & \dfrac{d\gamma_1}{d\beta_2} & \cdots & \dfrac{d\gamma_1}{d\beta_n} \\ \dfrac{d\gamma_2}{d\beta_1} & \dfrac{d\gamma_2}{d\beta_2} - e^{\alpha T} & \cdots & \dfrac{d\gamma_2}{d\beta_n} \\ \vdots & \vdots & \ddots & \vdots \\ \dfrac{d\gamma_n}{d\beta_1} & \dfrac{d\gamma_n}{d\beta_2} & \cdots & \dfrac{d\gamma_n}{d\beta_n} - e^{\alpha T} \end{vmatrix} = 0.$$

The left-hand side of this equation is a mapping in α; additionally, according to Theorem III, Sect. 2 in Chap. 1, the γ are expandable in powers of μ and β (see Sect. 1); furthermore according to Sect. 1, the β can themselves be expanded in powers of μ. Based on that, the γ and the determinant that I just wrote can themselves be expanded in powers of μ. From that, it follows that the exponents α are given as functions of μ by an equation

$$G(\alpha, \mu) = 0$$

whose left-hand side is a mapping in α and μ.

If for $\mu = 0$, all the exponents α were distinct from each other, the equation $G = 0$ would only have single roots for $\mu = 0$, and from that one would conclude that the α could be expanded in powers of μ (Theorem IV, Sect. 2 in Chap. 1).

But it is not that way; we just saw in fact that for $\mu = 0$, all the α are zero.

Continue with the notation from Sect. 3; by assuming only three degrees of freedom, our equation could be written as

$$0 = G(\alpha, \mu)$$

$$= \begin{vmatrix} \dfrac{d\Delta\alpha_1}{d\delta\alpha_1} + 1 - e^{\alpha T} & \dfrac{d\Delta\alpha_1}{d\delta\alpha_2} & \dfrac{d\Delta\alpha_1}{d\delta\alpha_3} & \dfrac{d\Delta\alpha_1}{d\delta\varpi_1} & \dfrac{d\Delta\alpha_1}{d\delta\varpi_2} & \dfrac{d\Delta\alpha_1}{d\delta\varpi_3} \\ \dfrac{d\Delta\alpha_2}{d\delta\alpha_1} & \dfrac{d\Delta\alpha_2}{d\delta\alpha_2} + 1 - e^{\alpha T} & \dfrac{d\Delta\alpha_2}{d\delta\alpha_3} & \dfrac{d\Delta\alpha_2}{d\delta\varpi_1} & \dfrac{d\Delta\alpha_2}{d\delta\varpi_2} & \dfrac{d\Delta\alpha_2}{d\delta\varpi_3} \\ \dfrac{d\Delta\alpha_3}{d\delta\alpha_1} & \dfrac{d\Delta\alpha_3}{d\delta\alpha_2} & \dfrac{d\Delta\alpha_3}{d\delta\alpha_3} + 1 - e^{\alpha T} & \dfrac{d\Delta\alpha_3}{d\delta\varpi_1} & \dfrac{d\Delta\alpha_3}{d\delta\varpi_2} & \dfrac{d\Delta\alpha_3}{d\delta\varpi_3} \\ \dfrac{d\Delta\varpi_1}{d\delta\alpha_1} & \dfrac{d\Delta\varpi_1}{d\delta\alpha_2} & \dfrac{d\Delta\varpi_1}{d\delta\alpha_3} & \dfrac{d\Delta\varpi_1}{d\delta\varpi_1} + 1 - e^{\alpha T} & \dfrac{d\Delta\varpi_1}{d\delta\varpi_2} & \dfrac{d\Delta\varpi_1}{d\delta\varpi_3} \\ \dfrac{d\Delta\varpi_2}{d\delta\alpha_1} & \dfrac{d\Delta\varpi_2}{d\delta\alpha_2} & \dfrac{d\Delta\varpi_2}{d\delta\alpha_3} & \dfrac{d\Delta\varpi_2}{d\delta\varpi_1} & \dfrac{d\Delta\varpi_2}{d\delta\varpi_2} + 1 - e^{\alpha T} & \dfrac{d\Delta\varpi_2}{d\delta\varpi_3} \\ \dfrac{d\Delta\varpi_3}{d\delta\alpha_1} & \dfrac{d\Delta\varpi_3}{d\delta\alpha_2} & \dfrac{d\Delta\varpi_3}{d\delta\alpha_3} & \dfrac{d\Delta\varpi_3}{d\delta\varpi_1} & \dfrac{d\Delta\varpi_3}{d\delta\varpi_2} & \dfrac{d\Delta\varpi_3}{d\delta\varpi_3} + 1 - e^{\alpha T} \end{vmatrix}$$

4 Calculation of the Characteristic Exponents

After doing that, I set

$$\alpha = \lambda\sqrt{\mu}.$$

I divide the first three lines of the determinant by $\sqrt{\mu}$; I next divide the last three columns by $\sqrt{\mu}$ (such that the determinant itself is finally divided by μ^3).

I next set $\mu = 0$.

I observe that according to what we saw in Sect. 3 Δa_1, Δa_2, and Δa_3 are divisible by μ. If I, therefore, consider the first element of the first row, after division by $\sqrt{\mu}$, this element will be written as

$$\frac{1}{\sqrt{\mu}}\frac{d\Delta\alpha_1}{d\delta\alpha_1} + \frac{1 - e^{\lambda T\sqrt{\mu}}}{\sqrt{\mu}},$$

and once we make $\mu = 0$ in it, it will become $-\lambda T$.

Similarly, the second element from the first row is written as

$$\frac{1}{\sqrt{\mu}}\frac{d\Delta\alpha_1}{d\delta\alpha_2},$$

and it approaches 0 as μ does.

Thus, once we have made $\mu = 0$, the first three elements of the first three rows will become zero except for the main diagonal elements which will become equal to $-\lambda T$.

Now, we consider the last three elements of the last three rows; they are written as

$$\frac{1}{\sqrt{\mu}}\frac{d\Delta\varpi_i}{d\delta\varpi_i} + \frac{1 - e^{\alpha T}}{\sqrt{\mu}} \quad \text{or} \quad \frac{1}{\sqrt{\mu}}\frac{d\Delta\varpi_k}{d\delta\varpi_i}$$

depending on whether they are on the main diagonal or not. Based on what we saw in Sect. 3, $\Delta\varpi_k$ is expandable in powers of μ, $\delta\alpha_i$, and $\delta\varpi_i$; additionally, for $\mu = 0$, $\Delta\varpi_i$ does not depend on the $\delta\varpi_i$. From that we conclude that $d\Delta\varpi_k/d\delta\varpi_i$ is divisible by μ.

Therefore, once we set $\mu = 0$, the last three elements of the last three rows will become equal to

$$-\lambda T \quad \text{or} \quad 0$$

depending upon whether or not they belong to the main diagonal.

Now consider the first three elements of the last three rows, $d\Delta\varpi_i/d\delta\alpha_k$. According to what we saw in Sect. 3, we have for $\mu = 0$:

$$\frac{d\Delta\varpi_i}{d\delta\alpha_k} = -T\frac{d^2 F_0}{dx_i dx_k}.$$

Now turn to the last three elements of the first three rows which are written as

$$\frac{1}{\mu}\frac{d\Delta\alpha_i}{d\delta\varpi_k}.$$

According to what we saw in Sect. 3, if one substitutes α_1, α_2, α_3, $n_1 t + \varpi_1$, $n_2 t + \varpi_2$, and $n_3 t + \varpi_3$ in F_1 in place of x_1, x_2, x_3, y_1, y_2, and y_3, it can be seen that F_1 becomes a periodic function of t with period T and if the average value of this periodic function is called ψ, then for $\mu = 0$,

$$\frac{\Delta\alpha_i}{\mu} = T\frac{d\psi}{d\varpi_i},$$

hence

$$\frac{1}{\mu}\frac{d\Delta\alpha_i}{d\delta\varpi_k} = T\frac{d^2\psi}{d\varpi_i d\varpi_k}.$$

The following identity needs to be noted:

$$n_1\frac{d\psi}{d\varpi_1} + n_2\frac{d\psi}{d\varpi_2} + n_3\frac{d\psi}{d\varpi_3} = 0.$$

We therefore see that for $\mu = 0$ it holds that

$$\frac{G(\lambda\sqrt{\mu},\mu)}{\mu^3 T^6} = \begin{vmatrix} -\lambda & 0 & 0 & \frac{d^2\psi}{d\varpi_1^2} & \frac{d^2\psi}{d\varpi_1 d\varpi_2} & \frac{d^2\psi}{d\varpi_1 d\varpi_3} \\ 0 & -\lambda & 0 & \frac{d^2\psi}{d\varpi_1 d\varpi_2} & \frac{d^2\psi}{d\varpi_2^2} & \frac{d^2\psi}{d\varpi_2 d\varpi_3} \\ 0 & 0 & -\lambda & \frac{d^2\psi}{d\varpi_1 d\varpi_3} & \frac{d^2\psi}{d\varpi_2 d\varpi_3} & \frac{d^2\psi}{d\varpi_3^2} \\ -\frac{d^2 F_0}{dx_1^2} & -\frac{d^2 F_0}{dx_1 dx_2} & -\frac{d^2 F_0}{dx_1 dx_3} & -\lambda & 0 & 0 \\ -\frac{d^2 F_0}{dx_1 dx_2} & -\frac{d^2 F_0}{dx_2^2} & -\frac{d^2 F_0}{dx_2 dx_3} & 0 & -\lambda & 0 \\ -\frac{d^2 F_0}{dx_1 dx_3} & -\frac{d^2 F_0}{dx_2 dx_3} & -\frac{d^2 F_0}{dx_3^2} & 0 & 0 & -\lambda \end{vmatrix}.$$

By setting this determinant equal to 0, a sixth degree equation in λ results; two of its roots are zero; we will not talk about them further because they relate to the two specific solutions of degenerate form that I talked about above. The other four solutions are in general distinct.

4 Calculation of the Characteristic Exponents

It results from Theorem IV of Sect. 2 in Chap. 1 that we can draw λ (and consequently α) from the equation

$$\frac{G(\lambda\sqrt{\mu},\mu)}{\mu^3 T^6} = 0$$

In the form of the series expanded in increasing powers of $\sqrt{\mu}$. I will add that λ is expandable in powers of μ and that the expansion of α contains only odd powers of $\sqrt{\mu}$. In fact the roots of the equation

$$G(\alpha,\mu) = 0$$

must be equal pairwise and have opposite sign (see Sect. 2). Therefore, α must change sign when I change $\sqrt{\mu}$ into $-\sqrt{\mu}$.

We will now prove that S_i and T_i can also be expanded in powers of $\sqrt{\mu}$.

S_i and T_i are in fact given by the following equations:

$$\begin{aligned}
\frac{dS_i}{dt} + \alpha S_i &= \sum \frac{d^2 F}{dy_i dx_k} S_k + \sum \frac{d^2 F}{dy_i dy_k} T_k, \\
\frac{dT_i}{dt} + \alpha T_i &= -\sum \frac{d^2 F}{dx_i dx_k} S_k - \sum \frac{d^2 F}{dx_i dy_k} T_k.
\end{aligned} \quad (2'')$$

Let β_i be the initial value of S_i and β'_i that of T_i; for an arbitrary value of t, the values of S_i and T_i are expandable in powers of μ, α, the β_i, and the β'_i according to Theorem III, Sect. 2 in Chap. 1. Additionally because of the linear form of the equations, these values will be linear and homogeneous functions of the β_i and the β'_i.

In order to use notations analogous to that of Sect. 1, let $\beta_i + \psi_i$ be the value of S_i and let $\beta'_i + \psi'_i$ be that of T_i for $t = T$. The condition for the solution to be periodic is

$$\psi_i = \psi'_i = 0.$$

The ψ_i and the ψ'_i are linear functions of the β_i and the β'_i; these equations are therefore linear in these quantities. In general these equations do not allow another solution besides

$$\beta_i = \beta'_i = 0,$$

such that Eq. (2'') has no periodic solution other than

$$S_i = T_i = 0.$$

But we know that if α is chosen so as to satisfy $G(\alpha, \mu) = 0$, then Eq. (2′) allows periodic solutions other than $S_i = T_i = 0$.

Consequently, the determinant of the linear equations $\psi_i = \psi'_i = 0$ is zero. From these equations, we can therefore draw the ratios

$$\frac{\beta_i}{\beta'_1} \quad \text{and} \quad \frac{\beta'_i}{\beta'_1}$$

in the form of series expanded in powers of α and μ.

Since β'_1 remains arbitrary, we shall agree to take $\beta'_1 = 1$ such that the initial value of T_1 is equal to 1. The β_i and the β'_i are then expanded in powers of α and μ; but as we have seen the S_i and the T_i are expandable in powers of α, μ, the β_i, and the β'_i, and additionally α are expandable in powers of $\sqrt{\mu}$.

Therefore, the S_i and the T_i are expandable in powers of $\sqrt{\mu}$.
Which was to be proved.

In particular, one will have

$$T_1 = T_1^0 + T_1^1 \sqrt{\mu} + T_1^2 \mu + \cdots.$$

Since, according to our assumption, β'_1, which is the initial value of T_1, must be equal to 1, whatever μ, for $t = 0$ it will hold

$$T_1^0 = 1, \quad 0 = T_1^1 = T_1^2 = \cdots = T_1^m = \cdots.$$

Since the existence of our series has been demonstrated, we are going to try to determine its coefficients.

We have

$$S_i^0 = 0, \quad T_i^0 = \eta_i^0$$

and

$$\begin{aligned}
\xi_i &= e^{\alpha t}\left(S_i^0 + S_i^1 \sqrt{\mu} + \cdots\right), \\
\frac{d\xi_i}{dt} &= e^{\alpha t}\left[\frac{dS_i^0}{dt} + \sqrt{\mu}\frac{dS_i^1}{dt} + \cdots + \alpha S_i^0 + \alpha\sqrt{\mu}S_i^1 + \cdots\right], \\
\eta_i &= e^{\alpha t}\left(T_i^0 + T_i^1 \sqrt{\mu} + \cdots\right), \\
\frac{d\eta_i}{dt} &= e^{\alpha t}\left[\frac{dT_i^0}{dt} + \sqrt{\mu}\frac{dT_i^1}{dt} + \cdots + \alpha T_i^0 + \alpha\sqrt{\mu}T_i^1 + \cdots\right].
\end{aligned} \quad (4)$$

We will additionally expand the second derivatives of F which enter Eq. (2) as coefficients by writing:

4 Calculation of the Characteristic Exponents

$$\frac{d^2 F}{dy_i dx_k} = A_{ik}^0 + \mu A_{ik}^2 + \mu^2 A_{ik}^4 + \cdots,$$

$$\frac{d^2 F}{dy_i dy_k} = B_{ik}^0 + \mu B_{ik}^2 + \mu^2 B_{ik}^4 + \cdots,$$

$$-\frac{d^2 F}{dx_i dx_k} = C_{ik}^0 + \mu C_{ik}^2 + \mu^2 C_{ik}^4 + \cdots,$$

$$-\frac{d^2 F}{dx_i dy_k} = D_{ik}^0 + \mu D_{ik}^2 + \mu^2 D_{ik}^4 + \cdots.$$

(5)

These expansions only contain integer powers of μ and unlike the expansions in Eq.(4) do not contain terms depending on $\sqrt{\mu}$.

Observe that

$$A_{ik}^0 = B_{ik}^0 = D_{ik}^0 = 0, \ C_{ik}^m = C_{ki}^m, \ B_{ik}^m = B_{ki}^m, \ A_{ik}^m = -D_{ki}^m. \quad (6)$$

In Eq. (2), we will substitute the values of Eqs. (4) and (5) in place of the ξ and η, their derivatives and the second derivatives of F. In the expressions of Eq. (4), I assume that α is expanded in powers of $\sqrt{\mu}$, except when this quantity α appears in an exponential, $e^{\alpha t}$.

We will next equate similar powers of $\sqrt{\mu}$ and will thus obtain a series of equations with which to successively determine

$$\alpha_1, \alpha_2, \alpha_3, \text{etc.}. \quad S_i^0, S_i^1, S_i^2, \ldots \quad T_i^0, T_i^1, T_i^2, \ldots.$$

I will only write the first of these equations resulting from successively setting equal the constant terms, the terms in $\sqrt{\mu}$, the terms in μ, etc. I will thus eliminate the factor $e^{\alpha t}$ which is found throughout.

First equate the terms in $\sqrt{\mu}$; it follows

$$\frac{dS_i^1}{dt} + \alpha_1 S_i^0 = \sum_k A_{ik}^0 S_k^1 + \sum_k B_{ik}^0 T_k^1,$$

$$\frac{dT_i^1}{dt} + \alpha_1 T_i^0 = \sum_k C_{ik}^0 S_k^1 + \sum_k D_{ik}^0 T_k^1.$$

(7)

Equating the terms in μ, it follows that

$$\frac{dS_i^2}{dt} + \alpha_1 S_i^1 + \alpha_2 S_i^0 = \sum_k \left(A_{ik}^0 S_k^2 + A_{ik}^2 S_k^0 + B_{ik}^0 T_k^2 + B_{ik}^2 T_k^0 \right), \quad (i = 1, 2, 3) \quad (8)$$

and additionally three analogous equations give the dT_i^2/dt.

If we now use the relations (6), Eq. (7) becomes

$$\frac{dS_i^1}{dt} = 0, \quad \frac{dT_i^1}{dt} + \alpha_i \eta_i^0 = \sum_k C_{ik}^0 S_k^1.$$

The first of these equations shows that S_1^1, S_2^1, and S_3^1 are constants. As for the second, it shows that dT_i^1/dt is a constant; however, since T_i^1 must be a periodic function, this constant must be zero, such that one has

$$\alpha_i \eta_i^0 = C_{i1}^0 S_1^1 + C_{i2}^0 S_2^1 + C_{i3}^0 S_3^1, \tag{9}$$

which establishes three relationships between the three constants η_i^0, the three constants S_i^1, and the unknown quantity α_1.

For its part, Eq. (8) will be written as

$$\frac{dS_i^2}{dt} + \alpha_1 S_i^1 = \sum_k B_{ik}^2 \eta_k^0.$$

Let the B_{ik}^2 be periodic functions of t; expand them according to Fourier series and let b_{ik} be the constant term from the B_{ik}^2. It will follow

$$\alpha_1 S_i^1 = \sum_k b_{ik} \eta_k^0$$

or, making use of Eq. (9), it follows

$$\alpha_1^2 S_i^1 = \sum_{k=1}^{3} b_{ik} \left(C_{k1}^0 S_1^1 + C_{k2}^0 S_2^1 + C_{k3}^0 S_3^1 \right). \tag{10}$$

By making $i = 1, 2, 3$ in Eq. (10), we will have three linear, homogeneous relationships between three constants S_i^1. By eliminating these three constants we will then have one third-degree equation which will determine λ_1^2.

If, for brevity, we set

$$e_{ik} = b_{i1} C_{1k}^0 + b_{i2} C_{2k}^0 + b_{i3} C_{3k}^0,$$

then the equation for this elimination will be written as

$$\begin{vmatrix} e_{11} - \alpha_1^2 & e_{12} & e_{13} \\ e_{21} & e_{22} - \alpha_1^2 & e_{23} \\ e_{31} & e_{32} & e_{33} - \alpha_1^2 \end{vmatrix} = 0. \tag{11}$$

4 Calculation of the Characteristic Exponents

It can also be written as

$$\begin{vmatrix} -\alpha_1 & 0 & 0 & C_{11}^0 & C_{12}^0 & C_{13}^0 \\ 0 & -\alpha_1 & 0 & C_{21}^0 & C_{22}^0 & C_{23}^0 \\ 0 & 0 & -\alpha_1 & C_{31}^0 & C_{32}^0 & C_{33}^0 \\ b_{11} & b_{12} & b_{13} & -\alpha_1 & 0 & 0 \\ b_{21} & b_{22} & b_{23} & 0 & -\alpha_1 & 0 \\ b_{31} & b_{32} & b_{33} & 0 & 0 & -\alpha_1 \end{vmatrix} = 0.$$

The determination of α_1 is the only part of the calculation which presents any difficulty.

The equations analogous to Eqs. (7) and (8) formed by equating the coefficients of similar powers of $\sqrt{\mu}$ in Eq. (2) can then be used without difficulty to determine the α_k, S_i^m, and T_i^m. We can therefore state the following result:

The characteristic exponents α are expandable according to increasing powers of $\sqrt{\mu}$.

Therefore, focusing our full attention on determining α_1 we will specifically study Eq. (11). We will first need to determine the quantities C_{ik}^0 and b_{ik}.

Obviously it holds that

$$C_{ik}^0 = -\frac{d^2 F_0}{dx_i^0 dx_k^0}$$

and

$$B_{ik}^2 = \frac{d^2 F_1}{dy_i^0 dy_k^0}$$

or

$$B_{ik}^2 = -\sum A m_i m_k \sin \omega \quad (\omega = m_1 y_1^0 + m_2 y_2^0 + m_3 y_3^0 + h)$$

and

$$b_{ik}^2 = -\sum_S A m_i m_k \sin \omega.$$

Using the conventions established in the previous section, the sum represented by the $\sum$ sign extends to all terms, whatever the integer values assigned to m_1, m_2, and m_3. The sum represented by the $\sum_S$ sign extends only to the terms for which

$$n_1 m_1 + n_2 m_2 + n_3 m_3 = 0.$$

Under the $\sum_s$ sign we consequently have

$$\omega = m_2\varpi_2 + m_3\varpi_3 + h.$$

This allows us to write

$$b_{ik} = \frac{d^2\psi}{d\varpi_i d\varpi_k} \quad \text{(for } i \text{ and } k = 2 \text{ or } 3\text{)}.$$

If one or both of the indices i and k are equal to 1, b_{ik} will be defined by the relation

$$n_1 b_{i1} + n_2 b_{i2} + n_3 b_{i3} = 0.$$

Using this last relation, we are going to transform Eq. (11) so as to demonstrate the existence of two zero roots and to reduce the equation to the fourth degree.

In fact, by a simple transformation of the determinant and dividing by α_1^2, I find

$$\begin{vmatrix} n_1 & n_2 & n_3 & 0 & 0 & 0 \\ 0 & -\alpha_1 & 0 & b_{23} & b_{22} & 0 \\ 0 & 0 & -\alpha_1 & b_{33} & b_{32} & 0 \\ C_{13}^0 & C_{23}^0 & C_{33}^0 & -\alpha_1 & 0 & n_3 \\ C_{12}^0 & C_{22}^0 & C_{32}^0 & 0 & -\alpha_1 & n_2 \\ C_{11}^0 & C_{21}^0 & C_{31}^0 & 0 & 0 & n_1 \end{vmatrix} = 0.$$

In the special case where there are only two degrees of freedom, this equation is written as

$$\begin{vmatrix} n_1 & n_2 & 0 & 0 \\ 0 & -\alpha_1 & \dfrac{d^2\psi}{d\varpi_2^2} & 0 \\ C_{12}^0 & C_{22}^0 & -\alpha_1 & n_2 \\ C_{11}^0 & C_{21}^0 & 0 & n_1 \end{vmatrix} = 0$$

or

$$n_1^2 \alpha_1^2 = \frac{d^2\psi}{d\varpi_2^2}\left(n_1^2 C_{22}^0 - 2n_1 n_2 C_{12}^0 + n_2^2 C_{11}^0\right).$$

The expression $n_1^2 C_{22}^0 - 2n_1 n_2 C_{12}^0 + n_2^2 C_{11}^0$ only depends on x_1^0 and x_2^0 or, if you prefer, on n_1 and n_2. When the two numbers n_1 and n_2, whose ratio must be

4 Calculation of the Characteristic Exponents

commensurable, are given, we can regard $n_1^2 C_{22}^0 - 2n_1 n_2 C_{12}^0 + n_2^2 C_{11}^0$ as a given constant. Then, the sign of α_1^2 depends only on that of $d^2\psi/d\varpi_2^2$.

When n_1 and n_2 are given, form the equation

$$\frac{d\psi}{d\varpi_2} = 0, \qquad (12)$$

which is Eq. (7) from the previous section. We saw in that section that for each root of this equation there is a corresponding periodic solution.

Consider the general case where Eq. (12) only has single roots; each of these roots then corresponds to a maximum or a minimum of ψ. But, since the function ψ is periodic, in each period it has at least one maximum and one minimum and precisely as many maxima as minima.

Hence, for the values of ϖ_2 corresponding to a minimum, $d^2\psi/d\varpi_2^2$ is positive; for the values corresponding to a maximum, this derivative is negative.

Therefore, Eq. (12) will have precisely as many roots for which this derivative will be positive as roots for which this derivative will be negative and consequently as many roots for which α_1^2 will be positive as roots for which α_1^2 will be negative.

This amounts to stating that there will be precisely as many stable periodic solutions as unstable solutions, giving this word the same meaning as in Sect. 2.

Thus, for each set of values of n_1 and n_2, there will correspond at least one stable periodic solution and one unstable periodic solution and precisely as many stable solutions as unstable solutions provided that μ is sufficiently small.

I will not examine here how these results would be extended to the case where Eq. (12) has multiple roots.

Here is how the calculation would need to be continued to higher orders.

Imagine that the quantities

$$\alpha_1, \alpha_2, \ldots \alpha_m$$

and the functions

$$S_i^0, S_i^1, \ldots S_i^m, \quad T_i^0, T_i^1, \ldots T_i^{m-1},$$

have been fully determined and that the functions S_i^{m+1} and T_i^m are known *up to a constant*. Assume that it is next proposed to calculate α_{m+1}, to complete the determination of the functions S_i^{m+1} and T_i^m and to then determine the functions S_i^{m+2} and T_i^{m+1} *up to a constant.*

Equations with the following form, analogous to Eqs. (7) and (8), can be obtained by equating similar powers of μ in Eq. (4):

$$-\frac{dT_i^{m+1}}{dt} + \sum_k C_{ik}^0 S_k^{m+1} - \alpha_1 T_i^m - \alpha_{m+1} T_i^0 \Bigg| = \text{known quantity},$$

$$-\frac{dS_i^{m+2}}{dt} + \sum_k B_{ik}^2 T_k^m - \alpha_1 S_i^{m+1} - \alpha_{m+1} S_i^1 = \text{known quantity}. \quad (i=1,2,3)$$

(13)

Both sides of these Eq. (13) are periodic functions of t. Equate the average value of these two sides. If we use $[u]$ to designate the average value of an arbitrary periodic function U and we observe that if U is periodic then it holds that

$$\left[\frac{dU}{dt}\right] = 0,$$

and if we recall that, since T_k^m is known up to a constant, $T_k^m - [T_k^m]$ and

$$[B_{ik}^2(T_k^m - [T_k^m])]$$

are known quantities, then we will get the following equations:

$$\sum_k C_{ik}^0 [S_k^{m+1}] - \alpha_1 [T_i^m] - \alpha_{m+1} T_i^0 = \text{known quantity},$$

$$\sum_k b_{ik}[T_k^m] - \alpha_1 [S_i^{m+1}] - \alpha_{m+1} S_i^1 = \text{known quantity}. \quad (i=1,2,3).$$

(14)

We are going to use Eq. (14) to calculate α_{m+1}, $[T_i^m]$, and $[S_i^{m+1}]$ and consequently complete the determination of the functions T_i^m and S_i^{m+1} which are still only known up to a constant.

If we add Eq. (14) after having multiplied them, respectively, by

$$S_1^1, S_2^1, S_3^1, T_1^0, T_2^0, T_3^0$$

it is found that

$$2\sum_i S_i^1 T_i^0 \alpha_{m+1} = \text{known quantity},$$

which determines α_{m+1}.

If in Eq. (14), α_{m+1} is replaced by the value found this way, then for determining six unknowns $[T_i^m]$ and $[S_i^{m+1}]$ there are six linear equations of which only five are independent.

With that stated, $[T_1^m]$ will be determined by the conditions that $[T_1^m]$ be zero for $t = 0$, consistent with the assumption made above, and the remaining five independent of Eq. (14) can be used to calculate the five other unknowns.

4 Calculation of the Characteristic Exponents

Equation (13) will allow us next to calculate dT_i^{m+1}/dt and dS_i^{m+2}/dt, and consequently determine the functions T_i^m and S_i^{m+1} up to a constant—and so on.

5 Asymptotic Solutions

Let

$$\frac{dx_i}{dt} = X_i \quad (i = 1, 2, \ldots n) \tag{1}$$

be n simultaneous differential equations. The X are functions of x and t.
  With respect to x, they are expandable in power series.
  With respect to t, they are periodic functions with period 2π.
  Let

$$x_1 = x_1^0, \quad x_2 = x_2^0, \quad \ldots \quad x_n = x_n^0.$$

be a specific periodic solution of these equations. The x_1^0 will be periodic functions of t with period 2π. Set

$$x_i = x_i^0 + \xi_i.$$

It will follow

$$\frac{d\xi_i}{dt} = \Xi_i. \tag{2}$$

The Ξ will be functions of ξ and t which are periodic with respect to t and expandable in powers of ξ, but there will no longer be any terms independent of ξ.
  If the ξ are very small and their squares neglected, then the equations reduce to

$$\frac{d\xi_i}{dt} = \frac{dX_i}{dx_1^0}\xi_1 + \frac{dX_i}{dx_2^0}\xi_2 + \cdots \frac{dX_i}{dx_n^0}\xi_n, \tag{3}$$

which are the perturbation equations of Eq. (1).
  They are linear and have periodic coefficients. The general form of their solution is known and we find:

$$\begin{aligned}
\xi_1 &= A_1 e^{\alpha_1 t}\varphi_{11} + A_2 e^{\alpha_2 t}\varphi_{21} + \cdots A_n e^{\alpha_n t}\varphi_{n1}, \\
\xi_2 &= A_1 e^{\alpha_1 t}\varphi_{12} + A_2 e^{\alpha_2 t}\varphi_{22} + \cdots A_n e^{\alpha_n t}\varphi_{n2}, \\
&\quad\vdots \\
\xi_n &= A_1 e^{\alpha_1 t}\varphi_{1n} + A_2 e^{\alpha_2 t}\varphi_{2n} + \cdots A_n e^{\alpha_n t}\varphi_{nn};
\end{aligned}$$

where the A are constants of integration, the α are fixed constants that are called characteristic exponents, and the φ are periodic functions of t.

If we then set

$$\begin{aligned}
\xi_1 &= \eta_1 \varphi_{11} + \eta_2 \varphi_{21} + \cdots \eta_n \varphi_{n1}, \\
\xi_2 &= \eta_1 \varphi_{12} + \eta_2 \varphi_{22} + \cdots \eta_n \varphi_{n2}, \\
&\vdots \\
\xi_n &= \eta_1 \varphi_{1n} + \eta_2 \varphi_{2n} + \cdots \eta_n \varphi_{nn},
\end{aligned}$$

Equation (2) will become

$$\frac{d\eta_i}{dt} = H_i, \tag{2'}$$

where the H_i are functions of t and η with the same form as the Ξ.

We can additionally write

$$\frac{d\eta_i}{dt} = H_i^1 + H_i^2 + \cdots H_i^n + \cdots ; \tag{2'}$$

where H_i^p represents the collection of terms of H_i which are of degree p in the η.

As for Eq. (3), they become

$$\frac{d\eta_i}{dt} = H_i^1 = \alpha_i \eta_i. \tag{3'}$$

Now we look for the form of the general solutions of Eqs. (2) and (2').

I state that we will need to find:

η_i = function expanded in powers of $A_1 e^{\alpha_1 t}, A_2 e^{\alpha_2 t}, \ldots A_n e^{\alpha_n t}$, whose coefficients are periodic functions of t. We can then write

$$\eta_i = \eta_i^1 + \eta_i^2 + \cdots \eta_i^p + \cdots, \tag{4'}$$

where η_i^p represents the collection of terms of η_i which are of degree p in the A.

We will replace the η_i with their values in H_i^p and we will find

$$H_i^p = H_i^{p,p} + H_i^{p,p+1} + \cdots H_i^{p,q} + \cdots,$$

where $H_i^{p,q}$ designates the terms which are of degree q in the A.

5 Asymptotic Solutions

We will then find:

$$\frac{d\eta_i^1}{dt} = \alpha_i \eta_i^1, \quad \eta_i^1 = A_i e^{\alpha_i t},$$

$$\frac{d\eta_i^2}{dt} - \alpha_i \eta_i^2 = H_i^{2,2}, \quad \frac{d\eta_i^3}{dt} - \alpha_i \eta_i^3 = H_i^{2,3} + H_i^{3,3},$$

$$\vdots$$

$$\frac{d\eta_i^q}{dt} - \alpha_i \eta_i^q = H_i^{2,q} + H_i^{3,q} + \cdots H_i^{q,q} = K_q.$$

With these equations it will be possible to successively calculate by iteration

$$\eta_i^2, \eta_i^3, \ldots \eta_i^q, \ldots$$

In fact, K_q depends only on the $\eta^1, \eta^2, \ldots \eta^{q-1}$. If we assume that these quantities had previously been calculated, we will be able to write K_q in the following form:

$$K_1 = \sum A_1^{\beta_1} A_2^{\beta_2} \ldots A_n^{\beta_n} e^{t(\alpha_1 \beta_1 + \alpha_2 \beta_2 + \cdots \alpha_n \beta_n)} \psi,$$

where the β are positive integers whose sum is q and ψ is a periodic function.

One can even write

$$\psi = \sum C e^{\gamma t \sqrt{-1}},$$

where C is a generally imaginary coefficient and γ is a positive or negative integer. For brevity, we will write

$$A_1^{\beta_1} A_2^{\beta_2} \cdots A_n^{\beta_n} = A^q, \quad \alpha_1 \beta_1 + \alpha_2 \beta_2 + \cdots \alpha_n \beta_n = \sum \alpha \beta,$$

and it will follow

$$\frac{d\eta_i^q}{dt} - \alpha_i \eta_i^q = \sum C A^q e^{t\left(\gamma\sqrt{-1} + \sum \alpha\beta\right)}.$$

Hence, this equation can be satisfied by setting

$$\eta_i^q = \sum \frac{C A^q e^{t\left(\gamma\sqrt{-1} + \sum \alpha\beta\right)}}{\gamma\sqrt{-1} + \sum \alpha\beta - \alpha_i}.$$

There could have been an exception in the case where it held

$$\gamma\sqrt{-1} + \sum \alpha\beta - \alpha_i = 0,$$

in which case it would add terms in t in the formulas. We will set this case aside, which does not occur in general.

We will now need to deal with the question of convergence of these series. The only difficulty, as is going to be seen, furthermore comes from the denominators

$$\gamma\sqrt{-1} + \sum \alpha\beta - \alpha_i. \tag{5}$$

This convergence is an immediate consequence of the results obtained in Sect. 3 in Chap. 1, but I prefer to give a direct proof.

Let us replace Eq. (2′) with the following:

$$\eta_i = \frac{1}{\varepsilon} A_i e^{\alpha_i t} + \overline{H_i^2} + \overline{H_i^3} + \cdots \overline{H_i^p} + \cdots. \tag{2″}$$

Now define $\overline{H_i^p}$. It can be seen without difficulty that H_i^p has the following form:

$$H_i^p = \sum C \eta_1^{\beta_1} \eta_2^{\beta_2} \ldots \eta_n^{\beta_n} e^{\gamma t \sqrt{-1}}.$$

C is an arbitrary constant, the β are positive integers whose sum is p, and γ is a positive or negative integer. We will then take

$$\overline{H_i^p} = \sum |C| \eta_1^{\beta_1} \eta_2^{\beta_2} \ldots \eta_n^{\beta_n}.$$

The resulting series will be convergent provided that the trigonometric series which define the periodic functions on which the H depend converge absolutely and uniformly; however, this will always be the case because these periodic functions are analytic. As for ε, it is a positive constant.

The η in the following can be drawn from Eq. (2″):

$$\eta_i = \sum M \varepsilon^{-\sum \beta} A_1^{\beta_1} A_2^{\beta_2} \ldots A_n^{\beta_n} e^{(\sum \alpha\beta)t}. \tag{4″}$$

Several terms could additionally have the same exponents β. By comparing with the series drawn from (2′) which are written as

$$\eta_i = \sum N \frac{A_1^{\beta_1} A_2^{\beta_2} \ldots A_n^{\beta_n}}{\Pi} e^{(\sum \alpha\beta + \gamma\sqrt{-1})t},$$

it can be observed that (1) M is positive, real, and larger than $|N|$, (2) Π designates the product of the denominators (5) ($q < \sum \beta$).

5 Asymptotic Solutions

Therefore, if the series (4″) converges and if none of the denominators (5) are smaller than ε, the series (4′) will also converge. Therefore, the convergence condition can be stated as follows.

The series converges,
if the expression

$$\gamma\sqrt{-1} + \sum \alpha\beta - \alpha_i$$

cannot become smaller than any given quantity ε for positive integer values of β and (positive or negative) integer values of γ; meaning if neither of the two convex polygons, the first surrounding the α and $+\sqrt{-1}$ and the second the α and $-\sqrt{-1}$, contain the origin;

or if all the quantities α have real parts with the same sign and if none of them have a zero real part.

What will we do if that is not so?

Assume, for example, that k of the quantities α have a positive real part and that $n - k$ have a negative or zero real part. It will then happen that the series (4′) will remain convergent if the constants A, which correspond to an α whose real part is negative or zero, cancel in it such that these series will no longer give the general solution of the proposed equations but a solution containing only k arbitrary constants.

If we assume that the given equations are among the equations of dynamics, we have seen that n is even and that the α are equal pairwise and of opposite side.

Then if k of them have a positive real part, k will have a negative real part and $n - 2k$ will have a zero real part. By first taking the α that have a positive real part, a specific solution is obtained containing k arbitrary constants; a second solution is obtained by taking the α that have a negative real part.

In the case where none of the α have a zero real part and in particular if all the α are real, then we also have

$$k = \frac{n}{2}.$$

We are now going to put ourselves in a very specific case. First assume $n = 2$, such that Eq. (1) reduce to

$$\frac{dx_1}{dt} = X_1, \quad \frac{dx_2}{dt} = X_2.$$

Additionally, assume that

$$\frac{dX_1}{dx_1} + \frac{dX_2}{dx_2} = 0. \tag{6}$$

The state of the system then depends on three quantities x_1, x_2, and t; the state can therefore be represented by the position of a point in space; in order to make the ideas more solid here is the mode of representation that can be adopted:

The rectangular components of the representative point will be

$$e^{x_1}\cos t, \quad e^{x_1}\sin t \quad \text{and} \quad x_2.$$

Such that

(1) Any set of values of the three quantities x_1, x_2, and t will correspond a point from the space;
(2) Any point from the space will correspond to a single set of values x_1, x_2, $\cos t$, and $\sin t$ and consequently a single state of the system, if two states which only differ because t has increased some number of periods 2π are not considered to be distinct;
(3) If t is varied (with x_1 and x_2 remaining constant), then the representative point describes a circumference;
(4) The condition $x_1 = x_2 = 0$, corresponds to the circle $z = 0$, $x^2 + y^2 = 1$;
(5) The condition $x_1 = -\infty$ corresponds to the z-axis.

To any solution of Eq. (1), they will correspond a curve described by the representative point. If the solution is periodic, this curve is closed.

We will therefore consider a closed curve C corresponding to a periodic solution.

Let us form Eqs. (2), (3), (2′), and (3′) relative to this periodic solution and let us imagine that the corresponding quantities α are being calculated.

The number these quantities is two, and because of the relation (6) they are equal and of opposite sign.

Two cases can come up: either their square is negative and the periodic solution is stable; or else their square is positive and the solution is unstable.

Place ourselves in the second case and call $+\alpha$ and $-\alpha$ the two values of the exponent α; we can assume then that α is real and positive.

With that said, the series (4′) are expanded according to increasing powers of $Ae^{\alpha t}$ and $Be^{-\alpha t}$; but they will not be convergent if A and B are simultaneously part of the series; the series will, however, become convergent if we make either $A = 0$, or $B = 0$.

First make $A = 0$; then the η will be expanded in powers of $Be^{-\alpha t}$; therefore, if t grows indefinitely, η_1 and η_2 tend simultaneously toward 0. The corresponding solutions can be called *asymptotic solutions*; because for $t = +\infty$, the η and consequently the ξ approach 0, which means that the asymptotic solution asymptotically approaches the periodic solution considered.

If, similarly, we make $B = 0$, the η are expanded in powers of $Ae^{\alpha t}$; they therefore approach 0 when t approaches $-\infty$. These are again asymptotic solutions.

There are therefore two series of asymptotic solutions, the first corresponding to $t = +\infty$ and the second to $t = -\infty$. Each of them contains an arbitrary constant, the first B and the second A.

5 Asymptotic Solutions

Each of these asymptotic solutions will correspond to a series of curves asymptotically approaching the closed curve C and the curves can be called asymptotic. The family of these asymptotic curves will form an *asymptotic surface*. There will be two asymptotic surfaces with the first corresponding to $t = +\infty$ and the second to $t = -\infty$. These two surfaces will pass by the closed curve C.

Suppose that in Eq. (1), the X depend on a parameter μ and that the functions X are expandable in powers of this parameter.

Let us imagine that for $\mu = 0$, the characteristic exponents α are all distinct such that these exponents, being defined by the equation $G(\alpha, \mu) = 0$ from the previous section, are themselves expandable in powers of μ.

Finally suppose that, as we just stated it, all the constants A which correspond to an α whose real part is negative or zero have been set to zero.

The series (4′) which define the quantities η_i then depend on μ. I propose to establish that these series are expandable, not only according to the powers of $A_i e^{\alpha_i t}$, but also in powers of μ.

Let us consider the inverse of one of the denominators (5):

$$\left(\gamma\sqrt{-1} + \sum \alpha\beta - \alpha_i\right)^{-1}.$$

I state that this expression is expandable in powers of μ.

Let $\alpha_1, \alpha_2, \ldots \alpha_k$ be the k characteristic exponents whose real part is positive and that we have agreed to retain. Each of them are expandable in powers of μ. Let α_i^0 be the value of α_i for $\mu = 0$; we will be able to take μ_0 small enough in order that α differs as little as we wish from α_i^0 when $|\mu| < \mu_0$. Then let h be a positive quantity smaller than the smallest of the real parts of the k quantities $\alpha_1^0, \alpha_2^0, \ldots \alpha_k^0$; we will be able to take μ_0 small enough such that when $|\mu| < \mu_0$, the real part of the k exponents $\alpha_1, \alpha_2, \ldots \alpha_k$ will be larger than h.

The real part of $\gamma\sqrt{-1} + \sum \alpha\beta - \alpha_i$ will then be larger than h (if $\beta_i > 0$), such that we will have

$$\left|\gamma\sqrt{-1} + \sum \alpha\beta - \alpha_i\right| > h.$$

Thus if $|\mu| < \mu_0$, the function

$$\left(\gamma\sqrt{-1} + \sum \alpha\beta - \alpha_i\right)^{-1}$$

remains one-to-one, continuous, finite, and smaller in absolute value than $1/h$.

We conclude from this, according to a well-known theorem, that this function is expandable in powers of μ and that the coefficients of the expansion are smaller in absolute value than those the expansion of

$$\frac{1}{h\left(1-\frac{\mu}{\mu_0}\right)}.$$

It should be noted that the numbers h and μ_0 are independent of the integers β and γ.

There would be an exception in the case, where β_i were zero. The real part of the denominator (5) could then be smaller than h and even be negative. It is in fact equal to the real part of $\sum \alpha\beta$, which is positive, minus the real part of α_i, which is also positive and which could be larger than the real part of $\sum \alpha\beta$ if β_i were zero.

Assume that the real part of α_i remains smaller than some number h_1 so long as $|\mu| < \mu_0$. Then if

$$\sum \beta > \frac{h_1}{h} + 1 \tag{7}$$

the real part of (5) is certainly larger than h; there can therefore only be a difficulty for those of the denominators (5) for which the inequality (7) does not hold.

Now assume that the imaginary part of the quantities $\alpha_1, \alpha_2, \ldots \alpha_k$ always remain smaller in absolute value than some positive number h_2; if it then holds that

$$|\gamma| > h_2 \sum \beta + h, \tag{8}$$

then the imaginary part of (5) and consequently its modulus will again be larger than h; such that there can only be a difficulty for those denominators (5) for which neither of the inequalities (7) and (8) hold. But these denominators which do not satisfy any of these inequalities are *finite in number*.

According to an assumption that we made above, none of them become zero for the values of μ that we are considering; we can therefore take h and μ_0 small enough such that the absolute value of any one among them remains larger than h when $|\mu|$ remains smaller than μ_0.

Then the inverse of an *arbitrary* denominator (5) is expandable in powers of μ and the coefficients of the expansion are smaller in absolute value than those of

$$\frac{1}{h\left(1-\frac{\mu}{\mu_0}\right)}.$$

Above, we wrote

$$H_i^p = \sum C \eta_1^{\beta_1} \eta_2^{\beta_2} \ldots \eta_n^{\beta_n} e^{\gamma t \sqrt{-1}}.$$

5 Asymptotic Solutions

According to our assumptions, C is expandable in powers of μ such that I can set

$$C = \sum E\mu^l, \quad H_i^p = \sum E\mu^l \eta_1^{\beta_1} \eta_2^{\beta_2} \ldots \eta_n^{\beta_n} e^{\gamma t\sqrt{-1}}.$$

Now return to Eq. (2″) by setting in them

$$\varepsilon = h\left(1 - \frac{\mu}{\mu_0}\right),$$

$$\overline{H_i^p} = \sum |E|\mu^l \eta_1^{\beta_1} \eta_2^{\beta_2} \ldots \eta_n^{\beta_n}.$$

The right-hand sides of Eq. (2″) will then be convergent series ordered in powers of μ, and $\eta_1, \eta_2, \ldots \eta_n$.

One can draw from them the η_i in the form of series (4″) which are convergent and ordered in powers of $\mu, A_1 e^{\alpha_1 t}, A_2 e^{\alpha_2 t}, \ldots A_k e^{\alpha_k t}$.

From Eq. (2′) we will additionally draw the η_i in the form of series (4′) which are ordered in powers of $\mu, A_1 e^{\alpha_1 t}, A_2 e^{\alpha_2 t}, \ldots A_k e^{\alpha_k t}, e^{+t\sqrt{-1}}, e^{-t\sqrt{-1}}$. Each of the terms from (4′) is smaller in absolute value than the corresponding terms from (4″) and since the series (4″) converges, the series (4′) will also converge.

6 Asymptotic Solutions of the Equations of Dynamics

Return to Eq. (1) from Sect. 3

$$\frac{dx_i}{dt} = \frac{dF}{dy_i} \quad \frac{dy_i}{dt} = -\frac{dF}{dx_i} \quad (i = 1, 2, \ldots n) \tag{1}$$

and the assumptions about them made at the beginning of Sect. 3.

In that section, we saw that these equations allow periodic solutions and from that we can conclude that provided that one of the corresponding characteristic exponents α is real, these equations also allow asymptotic solutions.

At the end of the previous section, we considered the case where in Eq. (1) of Sect. 5, the right-hand sides X_i are expandable in powers of μ, but where the characteristic exponents remain distinct from each other for $\mu = 0$.

In the case of the equations which we are now going to take up, meaning Eq. (1) from Sect. 3 to Sect. 6, the right-hand sides are still expandable in powers of μ; but all the characteristic exponents are zero for $\mu = 0$.

This results in a large number of major differences.

In the first place, the characteristic exponents α are not expandable in powers of μ, but in powers of $\sqrt{\mu}$ (see Sect. 4). Similarly, the functions that I called $\varphi_{i,k}$ at the beginning of Sect. 5 (and which, in the specific case of the equations of dynamics which concern us here, are none other than the functions S_i and T_i from Sect. 4) are expandable, not in powers of μ, but in powers of $\sqrt{\mu}$.

Then, in Eq. (2') from Sect. 5:

$$\frac{d\eta_i}{dt} = H_i$$

the right-hand side H_i is expandable in powers of η, $e^{+t\sqrt{-1}}$, $e^{-t\sqrt{-1}}$ and $\sqrt{\mu}$ (and not of μ).

From this we extract the η_i in the form of the series obtained in the previous section

$$\eta_i = \sum N \frac{A_1^{\beta_1} A_2^{\beta_2} \cdots A_n^{\beta_n}}{\Pi} e^{t[\sum \alpha\beta + \gamma\sqrt{-1}]}$$

and N and Π will be expanded in powers of $\sqrt{\mu}$.

Some number of questions then naturally comes up

(1) We know that N and Π are expandable in powers of $\sqrt{\mu}$; is this also true of the quotient N/Π?
(2) If that is so, then there exist series ordered in powers of $\sqrt{\mu}$, $A_i e^{\alpha_i t}$, $e^{+t\sqrt{-1}}$ and $e^{-t\sqrt{-1}}$ which *formally* satisfy the proposed equations; are these series convergent?
(3) If they are not convergent, what part of the calculation of the asymptotic solutions can be taken from them?

I propose to prove that N/Π is expandable in powers of $\sqrt{\mu}$ and that consequently there exist series ordered in powers of $\sqrt{\mu}$, $A_i e^{\alpha_i t}$, $e^{+t\sqrt{-1}}$, and $e^{-t\sqrt{-1}}$ which formally satisfy Eq. (1). One could be doubtful of it; in fact Π is the product of some number of denominators (5) from the previous section. All these denominators are expandable in powers of $\sqrt{\mu}$; but some of them, those for which γ is zero, approach 0 with $\sqrt{\mu}$. It can therefore happen that Π approaches 0 with μ and contains a factor of some power of $\sqrt{\mu}$. If then N does not contain this same power as a factor, then the quotient N/Π would be expanded according to increasing powers of $\sqrt{\mu}$, but the expansion would start with negative powers.

I state that it is not that way and that the expansion of N/Π contains only positive powers of $\sqrt{\mu}$.

Let us look at the mechanism by which these negative powers of $\sqrt{\mu}$ disappear. Set

$$A_i e^{\alpha_i t} = w_i$$

and let us consider the x and y as functions of variables t and w.

Before going farther it is important to make the following remark: among the $2n$ characteristic exponents α, two are zero and the others are equal pairwise and have opposite sign. We will only retain $n-1$ at most of these exponents by agreeing to

6 Asymptotic Solutions of the Equations of Dynamics

regard as zero the coefficients A_i and variables w_i which correspond to the $n+1$ rejected exponents. We will only retain those among the exponents whose real part is positive.

With that stated, Eq. (1) becomes:

$$\frac{dx_i}{dt} + \sum_k \alpha_k w_k \frac{dx_i}{dw_k} = \frac{dF}{dy_i}, \qquad (2)$$

$$\frac{dy_i}{dt} + \sum_k \alpha_k w_k \frac{dy_i}{dw_k} = -\frac{dF}{dx_i}. \qquad (3)$$

Starting from these equations, let us look to expand the x_i and the $y_i - n_i t$ according to increasing powers of $\sqrt{\mu}$ and w such that the coefficients are periodic functions of t.

We can write

$$\alpha_k = \alpha_k^1 \sqrt{\mu} + \alpha_k^2 \mu + \cdots = \sum \alpha_k^p \mu^{\frac{p}{2}}$$

because we saw in Sect. 4 how the characteristic exponents are expandable in powers of $\sqrt{\mu}$.

Also write

$$x_i = x_i^0 + x_i^1 \sqrt{\mu} + \cdots = \sum x_i^p \mu^{\frac{p}{2}},$$

$$y_i - n_i t = y_i^0 + y_i^1 \sqrt{\mu} + \cdots = \sum y_i^p \mu^{\frac{p}{2}},$$

where the x_i^p and y_i^p are functions of t and the w which are periodic in t and are expandable in powers of w.

If we substitute these values in the place of α_k, the x_i and the y_i in Eqs. (2) and (3), then both sides of these equations will be expanded in powers of $\sqrt{\mu}$.

Equate the coefficients of $\mu^{\frac{p+1}{2}}$ in both sides of Eq. (2) and the coefficients of $\mu^{p/2}$ in both sides of Eq. (3), the following equations result:

$$\frac{dx_i^{p+1}}{dt} + \sum_k \alpha_k^1 w_k \frac{dx_i^p}{dw_k} = Z_i^p + \sum_k \frac{d^2 F_1}{dy_i^0 dy_k^0} y_k^{p-1} \qquad (4)$$

$$\frac{dy_i^{p+1}}{dt} + \sum_k \alpha_k^1 w_k \frac{dy_i^{p-1}}{dw_k} = T_i^p - \sum_k \frac{d^2 F_0}{dx_i^0 dx_k^0} x_k^p,$$

where Z_i^p and T_i^p depend only on

$$x_i^0, x_i^1, \ldots x_i^{p-1},$$
$$y_i^0, y_i^1, \ldots y_i^{p-2}.$$

Let us agree, as we did above, to represent the average value of U by $[U]$, if U is a periodic function of t.

From Eq. (4), we can deduce the following:

$$\sum_k \alpha_k^1 w_k \frac{d[x_i^p]}{dw_k} = [Z_i^p] + \sum_k \left[\frac{d^2 F_1}{dy_i^0 dy_k^0} y_k^{p-1}\right], \qquad (5)$$

$$\sum_k \alpha_k^1 w_k \frac{d[y_i^{p-1}]}{dw_k} = [T_i^p] - \sum_k \frac{d^2 F_0}{dx_i^0 dx_k^0} [x_k^p].$$

Now assume that a prior calculation allowed us to know

$$x_i^0, x_i^1, \ldots x_i^{p-1}, x_i^p - [x_i^p],$$
$$y_i^0, y_i^1, \ldots y_i^{p-2}, y_i^{p-1} - [y_i^{p-1}].$$

Equation (5) will make it possible for us to calculate $[x_i^p]$ and $\left[y_i^{p-1}\right]$ and consequently x_i^p and y_i^{p-1}. With Eq. (4) we will then be able to determine

$$x_i^{p+1} - [x_i^{p+1}] \quad \text{and} \quad y_i^p - [y_i^p],$$

such that this procedure will provide us all the coefficients of the expansions of x_i and y_i by iteration.

The only difficulty is the determination of $[x_i^p]$ and $\left[y_i^{p-1}\right]$ by Eq. (5).

The functions $[x_i^p]$ and $\left[y_i^{p-1}\right]$ are expanded according to increasing powers of w and we are going to calculate the various terms of these expansions starting with the lowest order terms.

In order to do that we are going to return to the notation from Sect. 4, meaning we are going to set

$$-\frac{d^2 F_0}{dx_i^0 dx_k^0} = C_{ik}^0 \quad \text{and} \quad \frac{d^2 F_1}{dy_i^0 dy_k^0} = b_{ik}$$

(for the zero values of w).

6 Asymptotic Solutions of the Equations of Dynamics

If we then call ξ_i and η_i the coefficients of

$$w_1^{m_1} w_2^{m_2} \ldots w_{n-1}^{m_{n-1}}$$

in $[x_i^p]$ and $\left[y_i^{p-1}\right]$, we will have the following equations in order to determine these coefficients:

$$\sum_k b_{ik} \eta_k - S\xi_i = \lambda_i, \qquad (6)$$

$$\sum_k C_{ik}^0 \xi_k - S\eta_i = \mu_i.$$

In Eq. (6) λ_i and μ_i are known quantities because they only depend on

$$x_i^0, x_i^1, \ldots x_i^{p-1}, x_i^p - [x_i^p],$$
$$y_i^0, y_i^1, \ldots y_i^{p-2}, y_i^{p-1} - \left[y_i^{p-1}\right]$$

or on the terms of $[x_i^p]$ and $\left[y_i^{p-1}\right]$ whose degree in w is smaller than

$$m_1 + m_2 + \cdots m_{n-1}.$$

Additionally, for brevity we have set

$$S = m_1 \alpha_1^1 + m_2 \alpha_2^1 + \cdots m_{n-1} \alpha_{n-1}^1.$$

We therefore have a system of linear equations for the calculation of the coefficients ξ_i and η_i. The only difficulty would be if the determinant of these equations were zero; however, this determinant is equal to

$$S^2 \left[S^2 - \left(\alpha_1^1\right)^2\right] \left[S^2 - \left(\alpha_2^1\right)^2\right] \cdots \left[S^2 - \left(\alpha_{n-1}^1\right)^2\right].$$

It can only become zero for

$$S = 0, \quad S = \pm \alpha_i^1,$$

meaning for

$$m_1 + m_2 + \cdots m_{n-1} = 0 \text{ or } 1.$$

Therefore, there could only be difficulty in the calculation of the terms of degree 0 or 1 in w.

But we do not have to go back over the calculation of these terms; in fact, we learned how to calculate the terms independent of the w in Sect. 3 and the coefficients of

$$w_1, w_2, \ldots w_{n-1}$$

in Sect. 4.

The terms independent of the w are in fact nothing other than the series (8) from Sect. 3 and the coefficients of

$$w_1, w_2, \ldots w_{n-1}$$

are nothing other than the series S_i and T_i from Sect. 4.

All that is left is for me to say a word about the first approximations.

We will give to x_i^0 constant values which are nothing other than those that we designated thus in Sect. 3.

We will then have the following equations:

$$\frac{dy_i^0}{dt} = 0, \quad \frac{dx_i^1}{dt} = 0, \quad \frac{dy_i^1}{dt} + \sum_k \alpha_k^1 w_k \frac{dy_i^0}{dw_k} = -\sum_k \frac{d^2 F_0}{dx_i^0 dx_k^0} x_k^1, \tag{7}$$

$$\frac{dx_i^2}{dt} + \sum_k \alpha_k^1 w_k \frac{dx_i^1}{dw_k} = \frac{dF_1}{dy_i^0}.$$

In F_0, which only depends on the x_i, these quantities must be replaced by x_i^0. In F_1, the x_i are replaced by the x_i^0 and the y_i by $n_i t$. F_1 then becomes a periodic function of t whose period is T. As in Sects. 3 and 4, we will designate the average value of this periodic function F_1 by ψ; ψ is then a periodic function with period 2π in the y_i^0.

The first two Eq. (7) show that the y_i^0 and the x_i^1 depend only on the w. By equating the average values of the two sides in the last two of Eq. (7), it follows that

$$\sum \alpha_k^1 w_k \frac{dy_i^0}{dw_k} = \sum C_{ik}^0 x_k^1, \tag{8}$$

$$\sum \alpha_k^1 w_k \frac{dx_i^1}{dw_k} = \frac{d\psi}{dy_i^0}.$$

Equation (8) must serve to determine the y_i^0 and the x_i^1 as a function of the w. Can these equations be satisfied by substituting series expanded in powers of w in place of the y_i^0 and the x_i^1?

6 Asymptotic Solutions of the Equations of Dynamics

In order for us to find out, let us consider the following differential equations:

$$\frac{dy_i^0}{dt} = \sum C_{ik}^0 x_k^1, \tag{9}$$

$$\frac{dx_i^1}{dt} = \frac{d\psi}{dy_i^0}.$$

These differential equations where the unknown functions are the y_i^0 and the x_i^1 allow a periodic solution

$$x_i^1 = 0, \quad y_i^0 = \varpi_i,$$

where ϖ_i is the quantity designated that way in Sect. 3.

The characteristic exponents associated with this periodic solution are precisely the quantities α_k^1. Among these quantities we agreed to only retain those whose real part is positive. Equation (9) allows a family of asymptotic solutions and it is easy to see that these solutions come in the form of series expanded in powers of w. These series will then satisfy Eq. (8). These equations can therefore be solved.

Since the x_i^1 and the y_i^0 are thus determined, the remainder of the calculation, as we have seen, does not present any difficulty. There therefore exist series ordered by powers of $\sqrt{\mu}$, the w and $e^{\pm t\sqrt{-1}}$ and which formally satisfy Eq. (1).

This proves that the expansion of N/Π never begins with a negative power of $\sqrt{\mu}$.

Unfortunately, the series that thus result are not convergent.

In fact, let

$$\frac{1}{\gamma\sqrt{-1} + \sum \alpha\beta - \alpha_i}.$$

If γ is not zero, this expression is expandable in powers of $\sqrt{\mu}$; but the radius of convergence of the resulting series approaches 0 when $\gamma/\sum \beta$ approaches 0.

If therefore the various quantities $1/\Pi$ are expanded in powers of $\sqrt{\mu}$, one could always find among these quantities infinitely many for which the radius of convergence of the expansion is as small as one wants.

One could still hope, however, unlikely as it might seem, that it is not the same for the expansions of various quantities N/Π; but we will see rigorously in the following that it is not so in general; this faint hope must be given up and it has to be concluded that the series that we just formed diverge.

But even though they are divergent cannot we get something from them?

First, consider the following series which is simpler than those that we have seen:

$$F(w, \mu) = \sum_n \frac{w^n}{1 + n\mu}.$$

This series converges uniformly when μ remains positive and when w remains smaller in absolute value than some positive number w_o smaller than 1. Similarly the series

$$\frac{1}{p!} \frac{d^p F(w, \mu)}{d\mu^p} = \pm \sum \frac{n^{p-1} w^n}{(1 + n\mu)^p}$$

converges uniformly.

Now, if we try to expand $F(w, \mu)$ in powers of μ, the series to which this leads

$$\sum w^n (-n)^p \mu^p \tag{10}$$

does not converge. If, in this series, the terms where the exponent of μ is greater than p are neglected, a certain function results:

$$\Phi_p(w, \mu).$$

It is easy to see that the expression:

$$\frac{F(w, \mu) - \Phi_p(w, \mu)}{\mu^p}$$

approaches 0 when μ approaches 0 by positive values, such that the series (10) asymptotically represents the function $F(w, \mu)$ for small values of μ in the same way that the Stirling series asymptotically represents the Euler Gamma Function for large values of x.

The divergent series that we have come to form in this section are entirely analogous to the series (10).

In fact let us consider one of the series:

$$\sum \frac{N}{\Pi} w_1^{\beta_1} w_2^{\beta_2} \ldots w_k^{\beta_k} e^{\gamma t \sqrt{-1}} = F(\sqrt{\mu}, w_1, w_2, \ldots w_k, t) \tag{10'}$$

and

$$\sum w_1^{\beta_1} w_2^{\beta_2} \ldots w_k^{\beta_k} e^{\gamma t \sqrt{-1}} \frac{d^p \left(\frac{N}{\Pi}\right)}{(d\sqrt{\mu})^p} = \frac{d^p F}{(d\sqrt{\mu})^p};$$

these series are uniformly convergent provided that the w remain smaller in absolute value than some bounds and that $\sqrt{\mu}$ remains real.

6 Asymptotic Solutions of the Equations of Dynamics

If N/Π is expanded in powers of $\sqrt{\mu}$, the series (10') are divergent as we stated. Let us assume that the terms in the expansion where the exponent of $\sqrt{\mu}$ is greater than p are neglected, a certain function will result

$$\Phi_p(\sqrt{\mu}, w_1, w_2, \ldots w_k, t),$$

which is expandable in powers of w, $e^{\pm t\sqrt{-1}}$ and which will be a polynomial of degree p in $\sqrt{\mu}$.

It can be seen that the expression

$$\frac{F - \Phi_p}{\sqrt{\mu^p}}$$

approaches 0 when μ approaches 0 by positive values, and does so however large p might be.

In fact, if the collection of terms from the expansion of N/Π where the exponent of $\sqrt{\mu}$ is at most equal to p is designated by H_p, then it holds

$$\frac{F - \Phi_p}{\sqrt{\mu^p}} = \sum \frac{1}{\sqrt{\mu^p}} \left(\frac{N}{\Pi} - H_p\right) w_1^{\beta_1} w_2^{\beta_2} \ldots w_k^{\beta_k} e^{\gamma t \sqrt{-1}}$$

and the series from the right-hand side is *uniformly* convergent and all its terms approach 0 when μ approaches 0.

It can then be stated that the series that we have obtained in the present Sect. 6 represent asymptotic solutions for small values of μ in the same way that the Stirling series represents the Euler Gamma Function.

It can be better understood then in the following manner: let us assume just two degrees of freedom to make the ideas more definite; then we will continue to keep just one of the quantities w and we can write our equations in the following form:

$$\frac{dx_i}{dt} + \alpha w \frac{dx_i}{dw} = \frac{dF}{dy_i}, \quad \frac{dy_i}{dt} + \alpha w \frac{dy_i}{dw} = -\frac{dF}{dx_i} \quad (i = 1, 2)$$

by eliminating the indices from α and w since they have become unnecessary.

We know that α is expandable according to odd powers of $\sqrt{\mu}$ and consequently α^2 is expandable in powers of μ; conversely μ is expandable in powers of α^2; we can replace μ by this expansion such that F will be expanded in powers of α^2. For $\alpha = 0$, F reduces to F_0 which only depends on x_1 and x_2.

Let

$$x_i = \varphi_i(t), \quad y_i = \psi_i(t)$$

be the periodic solution which will serve as our starting point. Let us set, as in Sect. 4,

$$x_i = \varphi_i(t) + \xi_i, \quad y_i = \psi_i(t) + \eta_i$$

our equations will become

$$\frac{d\xi_i}{dt} + \alpha w \frac{d\xi_i}{dw} = \Xi_i, \quad \frac{d\eta_i}{dt} + \alpha w \frac{d\eta_i}{dw} = H_i, \tag{11}$$

where Ξ_i and H_i are expanded in powers of the ξ_i, the η_i, α^2, and the coefficients are periodic functions of t.

For $\alpha = 0$, dF/dy_i and consequently Ξ_i become zero; therefore Ξ_i is divisible by α^2 and I can set

$$\Xi_i = \alpha^2 X_i + \alpha^2 X'_i,$$

where $\alpha^2 X_i$ represents the collection of terms of first degree in the ξ and the η, and $\alpha^2 X'_i$ represents the collection of terms of higher degree.

Similarly, when α is zero, dF/dx_i and consequently H_i now only depend on the ξ_i and not the η_i.

I can therefore set

$$H_i = Y_i + Y'_i + \alpha^2 Q_i + \alpha^2 Q'_i,$$

where $Y_i + \alpha^2 Q_i$ represents the collection of terms of first degree in the ξ and the η, while $Y'_i + \alpha^2 Q'_i$ represents the collection of terms of higher than first degree. I additionally assume that Y_i and Y'_i depend only on ξ_1 and ξ_2.

Let us set

$$\xi_1 = \alpha \zeta_1, \quad \xi_2 = \alpha \zeta_2,$$

where Y_i will become divisible by α and Y'_i by α^2 such that I will be able to set

$$Y_i + \alpha^2 Q_i = \alpha Z_i, \quad Y'_i + \alpha^2 Q'_i = \alpha^2 Z'_i$$

and such that our equations will become

$$\frac{d\zeta_i}{dt} + \alpha w \frac{d\zeta_i}{dw} = \alpha X_i + \alpha X'_i, \tag{12}$$

$$\frac{d\eta_i}{dt} + \alpha w \frac{d\eta_i}{dw} = \alpha Z_i + \alpha^2 Z'_i.$$

Let us consider the following equations:

6 Asymptotic Solutions of the Equations of Dynamics

$$\frac{d\zeta_i}{dt} = \alpha X_i, \tag{13}$$

$$\frac{d\eta_i}{dt} = \alpha Z_i.$$

These equations are linear in the unknowns ζ_i and η_i. They are not different from Eq. (2) from Sect. 4, except in that ξ_1 and ξ_2 there are replaced by $\alpha\zeta_1$ and $\alpha\zeta_2$ here. According to what we saw in Sect. 4, the equation which defines the characteristic exponents has four roots: one equal to $+\alpha$, the other to $-\alpha$ and the two others equal to zero.

For the first root, meaning the root $+\alpha$, there will correspond a solution of Eq. (2) from Sect. 4 which we learned to form in that Sect. 4 and that we wrote this way

$$\xi_i = e^{\alpha t} S_i, \quad \eta_i = e^{\alpha t} T_i.$$

Recall that S_i^0 is zero and that consequently S_i is divisible by α.

For the second root $-\alpha$, there will similarly correspond another solution to Eq. (2) and we will write it as

$$\xi_i = e^{-\alpha t} S_i', \quad \eta_i = e^{-\alpha t} T_i'.$$

Finally for the two zero roots, they will correspond two solutions to Eq. (2) that we will write

$$\xi_i = S_i'', \qquad \eta_i = T_i'',$$
$$\xi_i = S_i''' + \alpha t S_i'', \quad \eta_i = T_i''' + \alpha t T_i''.$$

T_i', T_i'', T_i''' and S_i', S_i'', and S_i''' are periodic functions of t, as are S_i and T_i.

Additionally S_i', S_i'', and S_i''' will be divisible by α as is S_i.

Let us then set

$$\alpha\zeta_1 = S_1\theta_1 + S_1'\theta_2 + S_1''\theta_3 + S_1'''\theta_4,$$
$$\alpha\zeta_2 = S_2\theta_1 + S_2'\theta_2 + S_2''\theta_3 + S_2'''\theta_4,$$
$$\eta_1 = T_1\theta_1 + T_1'\theta_2 + T_1''\theta_3 + T_1'''\theta_4,$$
$$\eta_2 = T_2\theta_1 + T_2'\theta_2 + T_2''\theta_3 + T_2'''\theta_4.$$

The functions θ_i thus defined will play a role analogous to those of the functions η_i from Sect. 5. Equation (12) then becomes

$$\frac{d\theta_1}{dt} + \alpha w \frac{d\theta_1}{dw} - \alpha\theta_1 = \alpha\Theta_1, \quad \frac{d\theta_2}{dt} + \alpha w \frac{d\theta_2}{dw} + \alpha\theta_2 = \alpha\Theta_2, \tag{14}$$

$$\frac{d\theta_3}{dt} + \alpha w \frac{d\theta_3}{dw} = \alpha \theta_4 + \alpha \Theta_3, \quad \frac{d\theta_4}{dt} + \alpha w \frac{d\theta_4}{dw} = \alpha \Theta_4,$$

where $\Theta_1, \Theta_2, \Theta_3$ and Θ_4 are functions expanded in powers of $\theta_1, \theta_2, \theta_3, \theta_4$ and α in which all the terms are at least second degree in the θ, and whose coefficients are periodic functions of t. Additionally, the θ must be periodic functions of t and the terms in $\theta_1, \theta_2, \theta_3$ and θ_4 of first degree in w must reduce to w, 0, 0, and 0.

These equations in Eq. (14) are analogous to Eq. (2′) from Sect. 5.

That stated, let Φ be a function that, similar to $\Theta_1, \Theta_2, \Theta_3$ and Θ_4, are expandable in powers of $\theta_1, \theta_2, \theta_3, \theta_4, \alpha, e^{+t\sqrt{-1}}$ and $e^{-t\sqrt{-1}}$ and that is such that each of its coefficients are real and positive and larger in absolute value than the coefficient of the corresponding term in $\Theta_1, \Theta_2, \Theta_3$ and Θ_4; all the terms from Φ are also, as are those from Θ_i, of at least second degree in the θ.

Observe that the number

$$\frac{n\sqrt{-1}}{\alpha} + p$$

(where n is a positive, negative or zero integer, and where p is a positive integer and at least equal to 1) is always larger than 1 in absolute value, whatever, furthermore, n, p and α are.

Let us form the equations:

$$\theta_1 = w + \Phi, \quad \theta_2 = \Phi, \quad \theta_3 = \theta_4 + \Phi, \quad \theta_4 = \Phi \quad (15)$$

which are analogous to Eq. (2″) from Sect. 5.

The θ can be drawn from Eq. (14) in the form of series ordered in powers of w and $e^{\pm t\sqrt{-1}}$ and which are analogous to the series Eq. (4′) from Sect. 5. The θ can be drawn from Eq. (15) in the form of series ordered in powers of the same variables and analogous to the series Eq. (4″) from Sect. 5. Each of terms from these latter series is positive and larger in absolute value than the corresponding terms from the first series; if therefore they converge, then so do the series drawn from Eq. (14).

Hence, it is easy to see that if a number w_0 independent of α can be found such that if $|w| < w_0$, then the series drawn from Eq. (15) converge.

From this it follows that the series, ordered in powers of w and drawn from Eq. (14), converge uniformly however small α is; and consequently, however small μ might be, as I stated above.

We now have the θ in the form of ordered series in powers of w and $e^{\pm t\sqrt{-1}}$; the coefficients are known functions of α. If each of these coefficients is expanded in powers of α, the θ expanded in powers of α will result. The resulting series are divergent, as we saw above; nonetheless, let

6 Asymptotic Solutions of the Equations of Dynamics

$$\theta_i = \theta_i^0 + \alpha \theta_i^1 + \alpha^2 \theta_i^2 + \cdots \alpha^p \theta_i^p + \cdots \qquad (16)$$

be these series.
Set

$$H_1 = \Theta_1 + \theta_1, \quad H_2 = \Theta_2 - \theta_2, \quad H_3 = \Theta_3 + \theta_4, \quad H_4 = \Theta_4.$$

Set

$$\theta_i = \theta_i^0 + \alpha \theta_i^1 + \alpha^2 \theta_i^2 + \cdots \alpha^p \theta_i^p + \alpha^p u_i \qquad (17)$$

by equating θ_i to the first $p + 1$ terms from the series (16) plus a remainder term $\alpha^p u_i$.

If the θ_i are replaced in H_i by their expansions (17), the H_i are expandable in powers of α and can be written as

$$H_i = \Theta_i^0 + \alpha \Theta_i^1 + \alpha^2 \Theta_i^2 + \cdots \alpha^{p-1} \Theta_i^{p-1} + \alpha^p U_i,$$

where the Θ_i^k are independent of α while U_i are expandable in powers of α.

One then has the equations

$$\frac{d\theta_i^0}{dt} = 0, \quad \frac{d\theta_i^1}{dt} + w\frac{d\theta_i^0}{dw} = \Theta_i^0,$$

$$\frac{d\theta_i^2}{dt} + w\frac{d\theta_i^1}{dw} = \Theta_i^1, \quad \cdots \quad \frac{d\theta_i^p}{dt} + w\frac{d\theta_i^{p-1}}{dw} = \Theta_i^{p-1} \qquad (18)$$

and next

$$\frac{du_i}{dt} + \alpha w \frac{du_i}{dw} + \alpha w \frac{d\theta_i^p}{dw} = \alpha U_i. \qquad (19)$$

This is the form of the function U_i; the quantities θ_i^k can be regarded as known functions of t and w, defined by Eq. (18) and by Eq. (20) that I will write in a little; while the u_i remain unknown functions. Then, U_i is a function expanded in powers of w, $e^{\pm t\sqrt{-1}}$, α and the u_i. Additionally, any term of qth degree in the u_i is of at least degree $p(q-1)$ in α.

Let U_i^0 be what U_i becomes when α and the u_i are set to zero; one will have

$$w\frac{d[\theta_i^p]}{dw} = [U_i^0]. \qquad (20)$$

I can next, by setting

$$U'_i = U_i - w \frac{d\theta_i^p}{dw}$$

and then

$$V_1 = U'_1 - u_1, \quad V_2 = U'_2 + u_2, \quad V_3 = U'_3 - u_4, \quad V_4 = U'_4,$$

put Eq. (19) into the following form:

$$\frac{du_1}{dt} + \alpha w \frac{du_1}{dw} - \alpha u_1 = \alpha V_1, \quad \frac{du_2}{dt} + \alpha w \frac{du_2}{dw} + \alpha u_2 = \alpha V_2, \quad (21)$$

$$\frac{du_3}{dt} + \alpha w \frac{du_3}{dw} - \alpha u_4 = \alpha V_3, \quad \frac{du_4}{dt} + \alpha w \frac{du_4}{dw} = \alpha V_4.$$

It can then be seen that the V_i contain only terms of at least second degree in w and the u_i.

In fact the θ_i are divisible by w and reduce to w or to 0 when the terms of higher degree than first degree in w are dropped from it. From this it first follows that θ_i^p is divisible by w^2. Also, the right-hand side of Eq. (17) will contain only terms of at least first degree in w and u_i. Therefore, Θ_i only contains terms of at least second degree in w and the u_i. It follows from this that the only terms of first degree which can remain in U_1, U_2, U_3 and U_4 reduce respectively to u_1, $-u_2$, u_4, and 0.

Furthermore, $w d\theta_i^p/dw$ is divisible by w^2; therefore, the V_i contain only terms of at least second degree,
Which was to be proved.

The u_i can be drawn from Eq. (21) in the form of series expanded in powers of w and $e^{\pm t\sqrt{-1}}$. By applying the same reasoning to these equations as to Eq. (4) it can be proved that these series converge when $|w| < w_0$ and that the convergence remains uniform, however, small α is.

The same applies to the series which represent du_i/dw, d^2u_i/dw^2, etc.

From that it follows that an upper bound can be assigned to u_i, du_i/dw, d^2u_i/dw^2, etc. independent of α provided that $|w| < w_0$.

But now I want to prove that that still takes place for all positive values of w. Let us look again at the equations:

$$\frac{du_i}{dt} + \alpha w \frac{du_i}{dw} = U'_i.$$

U'_i can be regarded as a series expanded in powers of α and the u_i and for which the coefficients are functions of t and w. I state that this series remains convergent whatever t and w might be provided that α and the u_i are small enough. In fact it can only stop converging if the function:

$$F(x_1, x_2, y_1, y_2)$$

were to stop being expandable in powers of α, the v_i and the v'_i when the x_i are replaced in them by

$$x_i^0 + \alpha x_i^1 + \alpha^2 x_i^2 + \cdots \alpha^{p+1} x_i^{p+1} + \alpha^{p+1} v_i$$

and the y_i by

$$n_i t + y_i^0 + \alpha y_i^1 + \alpha^2 y_i^2 + \cdots \alpha^p y_i^p + \alpha^p v_i,$$

or, which amounts to the same thing, if the function F were to stop being expandable in powers of $x_i - x_i^0$ and $y_i - n_i, t - y_i^0$ for an arbitrary value of t or w (meaning for an arbitrary set of value of t, y_1^0 and y_2^0). However, it is clear that it is not this way.

I can therefore always find a function Φ expandable in powers of α and the u_i, but whose coefficients are constants instead of being functions of t and w like those of U'_i; and additionally make it such that the coefficient of an arbitrary term of Φ is real, positive, and larger in absolute value than the corresponding coefficient of U'_i ($i = 1, 2, 3, 4$), at least for the values of t and w that I will need to consider.

I will add that, according to the specific form of the functions U'_i, I can find two real, positive numbers M and β such that the function Φ satisfies the condition that I just stated if I take

$$\Phi = \frac{M(1 + u_1 + u_2 + u_3 + u_4)}{1 - \beta\alpha - \beta\alpha^p(u_1 + u_2 + u_3 + u_4)}.$$

If I consider values of w which are positive and less than some bound W, then in order to satisfy this condition I will need to take the numbers M and β that much larger since W will be larger; but since W will be finite, the numbers M and β will themselves be finite.

Now let w_1 be a positive value of w smaller than w_0. Based on what we have seen above, it is possible for $w = w_1$ to assign an upper bound to u_1, u_2, u_3, and u_4; let u_0 be this bound and it will then hold

$$|u_i| < u_0 \quad \text{for} \quad w = w_1.$$

Now let u' be a function defined by the following conditions

$$\frac{du'}{dt} + \alpha w \frac{du'}{dw} = \frac{\alpha M(4u' + 1)}{1 - \beta\alpha - 4\beta\alpha^p u'},$$

$$u' = u_0 \quad \text{for} \quad w = w_1.$$

Then obviously we will have for all values of t and w:

$$|u_i| < u'. \quad (i = 1, 2, 3, 4)$$

From this, one finds without difficulty:

$$\frac{1 - \beta\alpha - \beta\alpha^p}{4M} \log\left(\frac{1 + 4u'}{1 + 4u_0}\right) - \frac{\beta\alpha^p}{M}(u' - u_0) = \log\left(\frac{w}{w_1}\right)$$

and for $\alpha = 0$, we find:

$$\frac{1 + 4u'}{1 + 4u_0} = \left(\frac{w}{w_1}\right)^{4M},$$

which proves that u' remains finite when α approaches 0.

From this we must conclude that the quantities u_i also remain finite when α approaches 0.

From this it follows that the series

$$\theta_i^0 + \alpha\theta_i^1 + \alpha^2\theta_i^2 + \cdots$$

represents the function θ_i *asymptotically* (meaning in the way that the Stirling series does) or in other words that the expression

$$\frac{\theta_i - \theta_i^0 - \alpha\theta_i^1 - \alpha^2\theta_i^2 - \cdots \alpha^{p-1}\theta_i^{p-1}}{\alpha^{p-1}}$$

approaches 0 with α. In fact this expression is equal to

$$\alpha(\theta_i^p + u_i)$$

and we have just seen that $\theta_i^p + u_i$ remains finite when α approaches 0.

But this is not all, I state that du_i/dw that remains finite when α approaches 0. We in fact have

$$\frac{d}{dt}\left(\frac{du_i}{dw}\right) + \alpha w \frac{d}{dw}\left(\frac{du_i}{dw}\right) + \alpha\left(\frac{du_i}{dw}\right) = \alpha \sum_k \frac{dU'_i}{du_k}\frac{du_k}{dw} + \alpha\frac{dU'_i}{dw}.$$

dU'_i/du_k and dU'_i/dw are functions of t, w and α and the u_i; but according to what we have just see, we can assign upper bounds to the u_i; we can therefore also assign upper bounds to the dU'_i/du_k and dU'_i/dw.

6 Asymptotic Solutions of the Equations of Dynamics

Let us assume for example that we have

$$\left|\frac{dU'_i}{du_k}\right| < A, \quad \left|\frac{dU'_i}{dw}\right| < B \quad \text{for} \quad (w < W),$$

where A and B are two positive numbers.

Additionally, we know that we can assign a bound to du_i/dw for $w = w_1$. Assume for example that we have

$$\left|\frac{du_i}{dw}\right| < u'_0 \quad \text{for} \quad w = w_1,$$

where u'_0 is a positive number. Next let u' be a function defined as follows:

$$\frac{du'}{dt} + \alpha w \frac{du'}{dw} = \alpha u'(4A + W) + \alpha B,$$

$$u' = u'_0 \quad \text{for} \quad w = w_1$$

Then it will obviously hold that

$$\left|\frac{du_i}{dw}\right| < u'.$$

It can now be seen without difficulty that u' only depends on w and satisfies the equation

$$w \frac{du'}{dw} = u'(4A + W) + B.$$

Therefore u' is finite; therefore du_i/dw remains finite when α approaches 0. Therefore it holds *asymptotically* (with this word understood to have the same meaning as above) that

$$\frac{d\theta_i}{dw} = \frac{d\theta_i^0}{dw} + \alpha \frac{d\theta_i^1}{dw} + \alpha^2 \frac{d\theta_i^2}{dw} + \cdots.$$

In the same way it could be proved that it holds asymptotically that

$$\frac{d\theta_i}{dt} = \frac{d\theta_i^0}{dt} + \alpha \frac{d\theta_i^1}{dt} + \alpha^2 \frac{d\theta_i^2}{dt} + \cdots, \quad \frac{d^2\theta_i}{dw^2} = \frac{d^2\theta_i^0}{dw^2} + \alpha \frac{d^2\theta_i^1}{dw^2} + \alpha^2 \frac{d^2\theta_i^2}{dw^2} + \cdots.$$

This is therefore the final conclusion that we come to:

The series:

$$x_i^0 + \sqrt{\mu}x_i^1 + \mu x_i^2 + \cdots, \quad \eta_i t + y_i^0 + \sqrt{\mu}y_i^1 + \mu y_i^2 + \cdots$$

defined in this section are divergent, but they enjoy the same property as the Stirling series such that it holds asymptotically:

$$x_i = x_i^0 + \sqrt{\mu}x_i^1 + \mu x_i^2 + \cdots,$$
$$y_i = \eta_i t + y_i^0 + \sqrt{\mu}y_i^1 + \mu y_i^2 + \cdots.$$

Additionally, if D is an arbitrary differentiation sign, meaning if we set

$$Df = \frac{d^{\lambda_0 + \lambda_1 + \cdots \lambda_k} f}{dt^{\lambda_0} dw_1^{\lambda_1} dw_2^{\lambda_2} \ldots dw_k^{\lambda_k}}$$

then one will have asymptotically

$$Dx_i = Dx_i^0 + \sqrt{\mu}Dx_i^1 + \mu Dx_i^2 + \cdots,$$
$$Dy_i = D(\eta_i t + y_i^0) + \sqrt{\mu}Dy_i^1 + \mu Dy_i^2 + \cdots.$$

For more on the study of series analogous to Stirling series, I refer the reader to Sect. 1 of a monograph that I published in *Acta Mathematica* (volume 8, page 295).

Part II
Equations of Dynamics and the N-Body Problem

Chapter 4
Study of the Case with Only Two Degrees of Freedom

1 Various Geometric Representations

Let us return to Eq. (1) of Sect. 3 in Chap. 3:

$$\frac{dx_1}{dt} = \frac{dF}{dy_1}, \quad \frac{dx_2}{dt} = \frac{dF}{dy_2},$$
$$\frac{dy_1}{dt} = -\frac{dF}{dx_1}, \quad \frac{dy_2}{dt} = -\frac{dF}{dx_2}. \tag{1}$$

We will limit ourselves to the simplest case which is the one where there are only two degrees of freedom; I do not need to concern myself in fact with the case where there is only one degree of freedom because the equations of dynamics then are easily and simply integrated.

We will therefore assume that the function F only depends on four variables x_1, x_2, y_1, y_2. We will additionally assume that this function is one-to-one with respect to these four variables and periodic with period 2π with respect to y_1 and y_2.

The state of the system is therefore defined by the four quantities x_1, x_2, y_1, y_2, but this state is unchanged when y_1 or y_2, increase by 2π or a multiple of 2π. In other words and to review the language of Chap. 1, x_1 and x_2 are linear variables, whereas y_1 and y_2 are angular variables.

We are familiar with an integral of these Eq. (2); it is the following:

$$F(x_1, x_2, y_1, y_2) = C, \tag{2}$$

where C designates the constant from energy conservation. If this constant is regarded as part of the givens of the problem, then the four quantities x and y are no longer independent; they are related by the relationship (2). In order to determine the solution of the system, it will therefore be sufficient to arbitrarily set one of these four quantities. Consequently, it becomes possible to represent the state of the system by the position of a point P in space.

It can also happen for various reasons that the four variables x and y are subject, not only to the equality (2), but to one or more inequalities:

$$\varphi_1(x_1, x_2, y_1, y_2) > 0, \quad \varphi_2(x_1, x_2, y_1, y_2) > 0. \tag{3}$$

To be concrete let us assume that the inequalities (3) are for example written:

$$b < x_1 < a,$$

and that the equality (2) is such that when x_1 satisfies these inequalities, the fourth variable x_2 can be drawn from the relation (2) as a one-to-one function of the other three x_1, y_1 and y_2.

We can then represent the state of the system by a point whose rectangular coordinates are:

$$X = \cos y_1 [1 + \cos y_2 (cx_1 + d)], \quad Y = \sin y_1 [1 + \cos y_2 (cx_1 + d)],$$
$$Z = \sin y_2 (cx_1 + d),$$

where c and d are two new positive constants such that

$$0 < cb + d; \quad ca + d < 1.$$

It is clear in fact there for any state of the system, meaning any set of values x_1, y_1 and y_2 satisfying the conditions:

$$b < x_1 < a, \quad 0 < y_1 < 2\pi, \quad 0 < y_2 < 2\pi,$$

there corresponds one and only one point in the space, included between the two tori:

$$\left(1 - \sqrt{X^2 + Y^2}\right)^2 + Z^2 = (cb + d)^2,$$
$$\left(1 - \sqrt{X^2 + Y^2}\right)^2 + Z^2 = (ca + d)^2. \tag{4}$$

And inversely, for each point from the space included between these two tori there corresponds one and only one set of values x_1, y_1 and y_2 satisfying the preceding inequalities.

It can happen that the inequalities (3) are no longer written $b < x_1 < a$, but that however these inequalities combined with relation (2) lead as a consequence to:

$$b < x_1 < a.$$

If additionally x_2 is again a one-to-one function of three variables, the same mode of geometric representation is still applicable.

1 Various Geometric Representations

We can put ourselves in a still more general case:

Assume that we can find an auxiliary variable ξ, having the following property. If x_1, x_2, y_1 and y_2 simultaneously satisfy the equality (2) and the inequalities (3) then x_1 and x_2 can be expressed as continuous functions of ξ, y_1 and y_2. Additionally, because of the inequalities (3), ξ cannot become infinite and still be included between some bounds such that as a consequence of (2) and (3) it holds:

$$b < \xi < a.$$

We can then fully define the state of the system by stating the three variables ξ, y_1 and y_2, and represent it by a point P whose rectangular coordinates are:

$$X = \cos y_1 [1 + \cos y_2(c\xi + d)], \quad Y = \sin y_1 [1 + \cos y_2(c\xi + d)],$$
$$Z = \sin y_2(c\xi + d)$$

with the conditions:

$$c \geq 0, \quad 0 < cb + d, \quad ca + d < 1.$$

It can then be seen, as in the previous case, that any state of the system corresponds to one and only one point in the space included between the two tori (4), and inversely that for each point between these two tori there cannot correspond more than one state of the system.

It can happen that for $x_1 = a$ (or more generally for $\xi = a$), the state of the system remains the same whatever the value taken on by y_2. We will see it in the remainder of the examples. This is the case in polar coordinates; in general to define the position of a point, the two coordinates ρ and ω must be given, but if it is assumed that $\rho = 0$, then one is always at the same point, specifically the pole, whatever the value of ω.

In this case the constants c and d will be chosen such that

$$ca + d = 0.$$

The second of the two tori (4) then reduces to a circle:

$$Z = 0, \quad X^2 + Y^2 = 1.$$

At each of the points of this circle, y_2 is indeterminate, but nonetheless, as for $\xi = a$ where the state of the system does not depend on y_2, one and only one state of the system corresponds to each point of the circle.

It can then be said that for any state of the system there corresponds one point from the space inside the first of the two tori (4) and inversely, one point inside of this torus can correspond to only one state of the system.

I will look at yet another case.

Let us imagine that because of the inequalities (3), ξ can take on all positive values, such that:

$$a = 0, \quad b = +\infty.$$

Let us assume that for $\xi = 0$, the state of the system does not depend on y_2 and that for as $\xi = \infty$, this state does not depend on y_1.

We can then represent the state by a point whose rectangular coordinates are:

$$X = \cos y_1 e^{\xi \cos y_2}, \quad Y = \sin y_1 e^{\xi \cos y_2}, \quad Z = \xi \sin y_2.$$

For $\xi = 0$, it follows (whatever y_2)

$$X = \cos y_1, \quad Y = \sin y_1, \quad Z = 0.$$

The representative point is found on the circle

$$X^2 + Y^2 = 1, \quad Z = 0$$

and its position does not depend on y_2; this does not present a problem because by assumption the state of the system for $\xi = 0$ no longer depends on y_2.

For $\xi = \infty$, provided that $\cos y_2$ is negative, one finds:

$$X = Y = 0, \quad Z = \sin y_2.$$

The representative point is then found on the Z-axis and its position does not depend on y_1, but for $\xi = \infty$, the state of the system does not depend on y_1 either.

The mode of representation adopted is therefore legitimate.

The preceding needs to be supported by some examples. I will only deal with three examples here.

The first of these examples is the most important because it is a specific case of the three-body problem. Let us imagine two bodies, the first with a large mass, the second with a finite, but very small, mass and assume that these two bodies describe around their mutual center of gravity a circumference of uniform motion. Let us next consider a third body of infinitesimal mass, such that its motion is perturbed by the attraction of the first two bodies, but that it cannot perturb the orbit of these first two bodies. Let us limit ourselves additionally to the case where this third body remains in the plane of the two circumferences described by the first two masses.

This is the case of a small planet moving under the influence of the Sun and Jupiter when the eccentricity of Jupiter and the inclination of the orbits are neglected.

This is again the case of the Moon moving under the influence of the Sun and the Earth when the eccentricity of the Earth's orbit and the inclination of the Moon's orbit to the ecliptic are neglected.

1 Various Geometric Representations

We will define the position of the third body by its osculating elements at a given moment and we will write the equations of motion by adapting the notations of M.F. Tisserand in his Note from the *Comptes rendus des séances de l'académie des sciences* of January 31, 1887 [vol. 104 pp. 259–65]:

$$\frac{dL}{dt} = \frac{dR}{dl}, \quad \frac{dl}{dt} = -\frac{dR}{dL}, \\ \frac{dG}{dt} = \frac{dR}{dg}, \quad \frac{dg}{dt} = -\frac{dR}{dG}. \tag{5}$$

I designate the osculator major axis, eccentricity, and mean motion of the third mass by a, e and n; I call the mean anomaly of this third mass l and the longitude of its perihelion g.

I then set:

$$L = \sqrt{a}, \quad G = \sqrt{a(1-e^2)}.$$

I then choose the units such that the gravitational constant is equal to 1, that the mean motion of the second mass is equal to 1 and that the longitude of this second mass is equal to t.

Under these conditions, the angle through which the distance between the latter two masses is seen from the first only differs from $l+g-t$ by a periodic function of l with period 2π.

The function R is the ordinary perturbing function increased by $\frac{1}{2a} = \frac{1}{2L^2}$. This function only depends on L, G, l and $l+g-t$; because the distance from the second mass to the first is constant and the distance from the third to the first only depends on L, G, and l. This function is additionally periodic with period 2π both with respect to l and with respect to $l+g-t$.

From that one concludes that:

$$\frac{dR}{dt} + \frac{dR}{dg} = 0$$

and that Eq. (5) allow as integral:

$$R + G = \text{const.}$$

We are now going to try to bring Eq. (5) into the form of Eq. (1). To do that, we only need to set:

$$x_1 = G, \quad x_2 = L, \\ y_1 = g - t, \quad y_2 = l,$$

$$F(x_1, x_2, y_1, y_2) = R + G,$$

and Eq. (5) take on the form:

$$\frac{dx_i}{dt} = \frac{dF}{dy_i}, \quad \frac{dy_i}{dt} = -\frac{dF}{dx_i}. \tag{1}$$

The function F depends on a very small parameter μ which is the mass of the second body and we can write:

$$F = F_0 + \mu F_1.$$

F is periodic with respect to y_1 and y_2 which are angular variables, whereas x_1 and x_2 are linear variables. If one makes $\mu = 0$, F reduces to F_0 and:

$$F_0 = \frac{1}{2a} + G = x_1 + \frac{1}{2x_2^2},$$

then only depends on linear variables.

It follows from the very definition of L and G as functions of a and e that one must have:

$$L^2 > G^2 \quad \text{or} \quad x_2^2 > x_1^2,$$

which shows that x_1 can vary from $-x_2$ to $+x_2$.

If it is assumed that $x_1 = +x_2$, then the eccentricity is zero; the result of this is that that perturbing function in the state of the system now only depend on the longitude difference of the small masses, meaning that:

$$l + g - t = y_1 + y_2.$$

From this it can be deduced that:

$$\frac{dF}{dy_1} = \frac{dF}{dy_2},$$

and hence:

$$\frac{d(x_1 - x_2)}{dt} = 0, \tag{6}$$

from which one would conclude (because the initial value of $x_1 - x_2$ is assumed to be zero) that x_1 must remain constantly equal to x_2; but for Eq. (1) that is only a singular solution and must be rejected. As it relates to the "specific" solutions that we need to retain, Eq. (6) simply means that when $x_1 - x_2$ reaches the value 0, then this value is a maximum which is additionally a consequence of the inequality $x_2^2 > x_1^2$.

If we now assume $x_2 = -x_1$, the eccentricity will again be zero, but the motion will be retrograde (it is every time that x_1 and x_2 do not have the same sign); then F and the state of the system will now only depend on the angle:

1 Various Geometric Representations

$$-l + g - t = y_1 - y_2,$$

which yields:

$$\frac{dF}{dy_1} + \frac{dF}{dy_2} = 0.$$

I am now going to deal with the following question:
Find a variable ξ such that if x_1, x_2, y_1, y_2 satisfy the equalities and inequalities (2) and (3) (which reduces to

$$F = C, \quad x_1^2 < x_2^2$$

in the case which we are considering) then these four quantities can be expressed as one-to-one functions of the ξ, y_1 and y_2.

I will first take up the question for the case where $\mu = 0$ and where

$$F = F_0 = \frac{1}{2x_2^2} + x_1.$$

Let us imagine a plane and in this plane of point whose coordinates are:

$$X = x_1 - c, \quad Y = x_2.$$

Then the equalities and inequalities (2) and (3) are written:

$$X + \frac{1}{2Y^2} = 0, \quad -Y < X + c < Y.$$

Let us build the curve:

$$X + \frac{1}{2Y^2} = 0$$

and the two straight lines

$$X + c = \pm Y.$$

These straight lines and this curve can be in two different arrangements shown by Figs. 1 and 2.

Each of the two figures are made up of two halves which are symmetric around the x-axis, but we have only shown the half which is above this axis. In the case of Fig. 1, the curve provides us two useful arcs, BC and DE, whereas the arcs AB and CD must be rejected because of the inequality $Y^2 > (X+c)^2$. In the case of Fig. 2, there is only one useful arc BC and the arc AB must be rejected.

Fig. 1 Two useful arcs, *BC* and *DE*

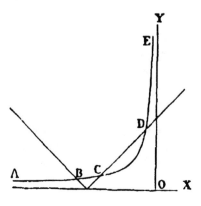

Fig. 2 Only one useful arc *BC* and the arc *AB* must be rejected

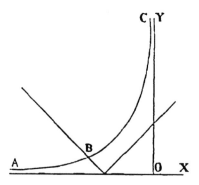

The switch from Fig. 1 and Fig. 2 occurs when the straight-line *CD* becomes tangent to the curve and the two points *C* and *D* merge. That takes place for:

$$C = \frac{3}{2}, \quad X = -\frac{1}{2}, \quad Y = 1.$$

In the following, we will assume we are located in the case from Fig. 1 and will consider only the useful arc *BC*; this is in fact the most interesting case from the perspective of applications.

Let us set:

$$\xi = \frac{x_2 - x_1}{x_2 + x_1} = \frac{L - G}{L + G};$$

it can be seen that ξ becomes zero at the point *C* and becomes infinite at the point *B* and that when the arc *BC* is traversed from *C* to *B*, ξ is seen to increase monotonically from 0 to $+\infty$. Therefore if ξ is given, the corresponding point of the arc *BC* will be completely determined, which amounts to stating that x_1 and x_2 are one-to-one functions of ξ.

Now what will happen if μ is not zero, but only very small?

1 Various Geometric Representations

Let us again make

$$\xi = \frac{x_2 - x_1}{x_2 + x_1}$$

and let us see if in light of the relationships[1]

$$F = C, \quad \xi > 0, \quad x_2 > 0, \tag{7}$$

x_1 and x_2 will again be one-to-one functions of ξ, y_1 and y_2. In order for it to stop being that way, the Jacobian determinant:

$$\frac{\partial(\xi, F)}{\partial(x_1, x_2)}$$

would have to be zero for a set of values satisfying conditions (7). However, that will not happen if μ is small enough and if C is sufficiently different from $3/2$.

In most applications, these conditions will be fulfilled; we can therefore take ξ as an independent variable; this variable will be essentially positive and x_1 and x_2 will be one-to-one functions of ξ, y_1, and y_2.

Just the same, to find the geometric representation that is easiest to work with, another change of variables is needed. Let us set:

$$x'_1 = x_1 + x_2, \quad x'_2 = x_1 - x_2, \quad y'_1 = \frac{1}{2}(y_1 + y_2), \quad y'_2 = \frac{1}{2}(y_1 - y_2).$$

After this change of variables, the equations will retain the canonical form:

$$\frac{dx'_1}{dt} = \frac{dF}{dy'_1}, \quad \frac{dy'_1}{dt} = -\frac{dF}{dx'_1}, \quad \frac{dx'_2}{dt} = \frac{dF}{dy'_2}, \quad \frac{dy'_2}{dt} = -\frac{dF}{dx'_2}.$$

It can be seen that y'_1 and y'_2 are again angular variables; in fact, when y'_1 or y'_2 increase by a multiple of 2π, y_1 and y_2 also increase by a multiple of 2π and consequently the state of the system does not change.

But there is more; when y'_1 and y'_2 simultaneously change into $y'_1 + \pi$ and $y'_2 + \pi$, y_2 does not change and y_1 increases by 2π. The state of the system is therefore unchanged.

Having established that, we will represent the state of the system by the spatial point whose rectangular coordinates are:

$$X = \cos y'_1 e^{\xi \cos y'_2}, \quad Y = \sin y'_1 e^{\xi \cos y'_2}, \quad Z = \xi \sin y'_2.$$

---

[1]The reason why I'm writing this last relationship can easily be seen: the arc BC is entirely above the X-axis, as can be seen in the figure, which leads to the inequality $x_2 > 0$; it is clear that this inequality will still be satisfied for sufficiently small values of μ.

For $\xi = 0$, the state of the system does not depend on y'_2 and the same is true for the representative point which is then on the circle

$$X^2 + Y^2 = 1, \quad Z = 0.$$

For $\xi = \infty$, the state of the system does not depend on y'_1 and it is the same for the representative point which is then on the Z-axis if $\cos y'_2$ is negative and at infinity if $\cos y'_2$ is positive.

There therefore corresponds to each spatial point one and only one state of the system; conversely, for each state of the system there corresponds, not one, but two spatial points and in fact two families of values (x'_1, x'_2, y'_1, y'_2) and $(x'_1, x'_2, y'_1 + \pi, y'_2 + \pi)$ corresponding to two different points in space, but one single state of the system.

Equation (1) allow the integral invariants:

$$\int (dx_1 dy_1 + dx_2 dy_2) = \int (dx'_1 dy'_1 + dx'_2 dy'_2)$$

and

$$\int dx_1 dy_1 dx_2 dy_2 = \int dx'_1 dy'_1 dx'_2 dy'_2.$$

If we transform this invariant by the rules presented in Sect. 3 of Chap. 2 we will see that:

$$\int \frac{x'^2_1 d\xi dy'_1 dy'_2}{x'_1 \frac{dF}{dx'_1} + x'_2 \frac{dF}{dx'_2}} = \int \frac{x'^2_1 dX dY dZ}{\left(x'_1 \frac{dF}{dx'_1} + x'_2 \frac{dF}{dx'_2}\right) \xi (X^2 + Y^2)}$$

is again an integral invariant.

Since ξ is mostly positive, the quantity under the integral sign has the same sign as:

$$x'_1 \frac{dF}{dx'_1} + x'_2 \frac{dF}{dx'_2} = x_1 \frac{dF}{dx_1} + x_2 \frac{dF}{dx_2}.$$

Hence for $\mu = 0$, we find that:

$$x_1 \frac{dF}{dx_1} + x_2 \frac{dF}{dx_2} = x_1 - \frac{1}{x_2^2}.$$

If we assume we find ourselves in the case from Fig. 1 and on the arc BC, we need to assume:

1 Various Geometric Representations

$$C > \tfrac{3}{2}, \quad x_1^2 < x_2^2, \quad 0 < x_2 < 1,$$

from which one concludes:

$$x_1 - \frac{1}{x_2^2} = 3x_1 - 2C < 3x_1 - 2 \times \frac{3}{2} = 3(x_1 - 1) < 0.$$

Thus $x_1 dF/dx_1 + x_2 dF/dx_2$ is always negative when μ is zero. It will still be the same when μ stops being zero, provided that C is sufficiently different from $3/2$.
Under these conditions the integral:

$$\int \frac{x_1'^2 \mathrm{d}X\mathrm{d}Y\mathrm{d}Z}{\left(-x_1' \dfrac{\mathrm{d}F}{\mathrm{d}x_1'} - x_2' \dfrac{\mathrm{d}F}{\mathrm{d}x_2'}\right) \xi(X^2 + Y^2)}$$

is a positive invariant.

For $\mu = 0$, Eq. (5) are easily solved as is known and one finds:

$$L = \text{const.}, \quad G = \text{const.}, \quad g = \text{const.}, \quad l = nt + \text{const.}$$

The resulting solutions are shown in the geometric representation mode adopted for some trajectories. These trajectories are closed whenever the mean motion n is a commensurable number. They are traced on trajectory surfaces which have

$$\xi = \text{const.}$$

for general equation and which are consequently closed surfaces of revolution analogous to tori.

Later, we will see how these results are changed when μ is no longer zero.

As a second example, let me go back to the equation that I already spoke about at the end of Sect. 3 in Chap. 3:

$$\frac{\mathrm{d}^2 \rho}{\mathrm{d}t^2} + n^2 \rho + m \rho^3 = \mu R,$$

where R is a function of ρ and t, mapping with respect to ρ and approaching zero with ρ, and periodic with respect to t. This equation can be written using the notation of the section referenced:

$$\frac{\mathrm{d}\rho}{\mathrm{d}t} = \frac{\mathrm{d}F}{\mathrm{d}\sigma}, \quad \frac{\mathrm{d}\sigma}{\mathrm{d}t} = -\frac{\mathrm{d}F}{\mathrm{d}\rho}, \quad \frac{\mathrm{d}\xi}{\mathrm{d}t} = \frac{\mathrm{d}F}{\mathrm{d}\eta}, \quad \frac{\mathrm{d}\eta}{\mathrm{d}t} = -\frac{\mathrm{d}F}{\mathrm{d}\xi}$$

with

$$\xi = t, \quad \frac{d\rho}{dt} = \sigma, \quad F = \frac{\sigma^2}{2} + \frac{n^2\rho^2}{2} + \frac{m\rho^4}{4} - \mu \int R(\rho, \xi) d\rho + \eta.$$

Let us set:

$$\sigma = \sqrt{n}\sqrt{2x_1} \cos y_1 \quad \rho = \frac{1}{\sqrt{n}}\sqrt{2x_1} \sin y_1$$

The equations will retain the canonical form of the equations of dynamics and the functions F will depend on two linear variables x_1 and η and two angular variables y_1 and ξ.

It can be easily seen that when the total energy constant C, and x_1, y_1 and ξ are given, then the fourth variable η is fully determined; in fact it holds:

$$\eta = C - nx_1 - \frac{m}{n^2} x_1^2 \sin^2 y_1 + \mu \int R(\rho, \xi) d\rho$$

For $x_1 = 0$, the state of the system does not depend on y_1. To represent this state, we can therefore adopt the point whose coordinates are:

$$X = \cos \xi e^{x_1 \cos y_1}, \quad Y = \sin \xi e^{x_1 \cos y_1}, \quad Z = x_1 \sin y_1.$$

To each state of the system there therefore corresponds one spatial point and conversely. An exception has to be made for the points at infinity and the points on the Z-axis which would give us $x_1 = \infty$ and consequently an illusory result.

As a third example let us consider a moving point subject to gravity moving on a frictionless surface in the area of a position of stable equilibrium.

For the origin, let us take the lowest point of the surface; for the xy-plane take the plane which is tangent and horizontal; and take for the x- and y-axis the axes of the indicatrix such that the equation of the surface can be written:

$$z = \frac{ax^2}{2} + \frac{by^2}{2} + \mu\varphi(x, y),$$

where $\varphi(x, y)$ is a set of terms of at least third degree in x and y and where μ is a very small coefficient.

By calling x' and y' the projections of the velocity on the x- and y-axis, we will then have:

$$F = \frac{x'^2}{2} + \frac{y'^2}{2} + gz,$$

$$\frac{dx}{dt} = \frac{dF}{dx'}, \quad \frac{dy}{dt} = \frac{dF}{dy'}, \quad \frac{dx'}{dt} = -\frac{dF}{dx}, \quad \frac{dy'}{dt} = -\frac{dF}{dy}.$$

1 Various Geometric Representations

Let us change variables by setting:

$$x = \frac{\sqrt{2x_1}}{\sqrt[4]{ga}}\cos y_1, \quad x' = \sqrt{2x_1}\sqrt[4]{ga}\sin y_1,$$

$$y = \frac{\sqrt{2x_2}}{\sqrt[4]{gb}}\cos y_2, \quad y' = \sqrt{2x_2}\sqrt[4]{gb}\sin y_2.$$

The differential equations will retain the canonical form of the equations of dynamics. The energy conservation equation is written:

$$\sqrt{ga}x_1 + \sqrt{gb}x_2 + \mu g \varphi(x_1, x_2, y_1, y_2) = C,$$

where φ designates the same function as above, but transformed by the change of variables. Because x_1 and x_2 are mainly positive (and the coefficients a and b as well), the energy conservation equation shows that these two quantities always remain smaller than some bound. According to the definition of the function φ, this function approaches zero with x_1 and x_2, and so do its first-order partial derivatives. From this, we conclude that since μ is very small, the function $\mu\varphi$ and its first-order derivatives can never exceed some very small upper bound. We can therefore write:

$$\left|\mu\frac{d\varphi}{dx_1}\right| < \sqrt{\frac{a}{g}}, \quad \left|\mu\frac{d\varphi}{dx_2}\right| < \sqrt{\frac{b}{g}}.$$

Now set $x_2 = \xi x_1$; the ratio ξ is mainly positive. The energy conservation equation becomes:

$$x_1\left(\sqrt{ga} + \sqrt{gb}\xi\right) + \mu g \varphi(x_1, \xi x_1, y_1, y_2) = C. \tag{9}$$

The derivative of the left-hand side of (9) with respect to x_1 is written:

$$\sqrt{ga} + \sqrt{gb}\xi + \mu g \frac{d\varphi}{dx_1} + \mu \xi g \frac{d\varphi}{dx_1}.$$

Because of the inequalities (8), this expression is always positive, which shows that x_1 can always be found from Eq. (9) as a one-to-one function of ξ, y_1 and y_2, and consequently the state of the system is fully defined by the three variables y_1, y_2 and ξ.

For $\xi = 0$, the state does not depend on y_1, for $\xi = \infty$, it does not depend on y_2. We will now represent this state by the point:

$$X = \cos y_2 e^{\xi \cos y_1}, \quad Y = \sin y_2 e^{\xi \cos y_1}, \quad Z = \xi \sin y_1.$$

To each point from this space there will therefore correspond one state of the system and inversely.

The preceding examples will suffice, I think, to help understand the importance of the problem which we are going to deal with in this chapter and the way in which the modes of geometric representation can be varied.

Chapter 5
Study of the Asymptotic Surfaces

1 Description of the Problem

Let us return to the equations of dynamics by assuming only two degrees of freedom and consequently four variables x_1, x_2, y_1, and y_2. According to what we have seen in Sect. 6 of Chap. 3, these equations allow some remarkable specific solutions which we called asymptotic. Each of these asymptotic solutions is represented in the system of geometric representation presented in the preceding section by certain trajectory curves. The family of these curves will lead to certain surfaces that we can call asymptotic surfaces and that we propose to study.

These asymptotic solutions can be put into the following form:

$$x_1 = \varphi_1(t,w), \quad x_2 = \varphi_2(t,w),$$
$$y_1 = n_1 t + \varphi_3(t,w), \quad y_2 = n_2 t + \varphi_4(t,w) \tag{1}$$

where w is equal to $Ae^{\alpha t}$, and A is an arbitrary constant. Furthermore φ_1, φ_2, φ_3, and φ_4 are (with respect to t, where w is regarded for a moment as a constant) periodic functions with period T, and $n_1 T$ and $n_2 T$ are multiples of 2π.

If t and w are eliminated from this Eq. (1), it follows that:

$$x_1 = f_1(y_1, y_2), \quad x_2 = f_2(y_1, y_2) \tag{2}$$

and these equations can be regarded as defining our asymptotic surfaces. We next saw that when trying to expand φ_1, φ_2, φ_3, and φ_4 in powers of $\sqrt{\mu}$, we arrived at series which are divergent, but that these series nonetheless asymptotically represent these functions when μ is very small.

Remember that I agreed to state that the series

$$A_0 + A_1 x + \cdots A_p x^p + \cdots$$

asymptotically represents the function $F(x)$ for very small x, when

$$\lim_{x \to 0} \frac{F(x) - A_0 - A_1 x - \cdots A_p x^p}{x^p} = 0.$$

In *Acta Mathematica*, volume 8, I studied the properties of divergent series which asymptotically represent certain functions and I recognized that the ordinary rules of calculation are applicable to these series. An equality, meaning an equality between a divergent series and a function that it represents asymptotically, can undergo all the ordinary calculation operations, except for differentiation.

Therefore let:

$$\sigma_1(t, w, \sqrt{\mu}), \quad \sigma_2(t, w, \sqrt{\mu}), \quad \sigma_3(t, w, \sqrt{\mu}), \quad \sigma_4(t, w, \sqrt{\mu})$$

be the divergent series ordered by powers of $\sqrt{\mu}$ which asymptotically represent φ_1, φ_2, φ_3, and φ_4.

We will then have the four asymptotic equalities:

$$\begin{aligned} x_1 &= \sigma_1(t, w, \sqrt{\mu}), & x_2 &= \sigma_2(t, w, \sqrt{\mu}), \\ y_1 &= n_1 t + \sigma_3(t, w, \sqrt{\mu}), & y_2 &= n_2 t + \sigma_4(t, w, \sqrt{\mu}). \end{aligned} \quad (3)$$

We will be able to eliminate t and w from these equalities using the ordinary rules of calculation and we will thus get two new asymptotic equalities:

$$x_1 = s_1(y_1, y_2, \sqrt{\mu}), \quad x_2 = s_2(y_1, y_2, \sqrt{\mu}), \quad (4)$$

where s_1 and s_2 are divergent series ordered by powers of $\sqrt{\mu}$ and for which the coefficients are functions of y_1 and y_2.

In general, differentiating an asymptotic equality is not allowed; but at the end of Sect. 6 in Chap. 3 we directly proved that in the specific case with which we are concerned the equalities (3) can be differentiated both with respect to t and with respect to w as many times as we want.

From this we can also conclude that differentiating the equalities (4) with respect to y_1 and y_2 as many times as we want is allowed.

We propose to study the asymptotic surfaces defined by Eq. (2). The functions $x_1 = f_1$ and $x_2 = f_2$ which enter into these equations will need to satisfy the equations:

$$\begin{aligned} \frac{dF}{dx_1}\frac{dx_1}{dy_1} + \frac{dF}{dx_2}\frac{dx_1}{dy_2} + \frac{dF}{dy_1} &= 0, \\ \frac{dF}{dx_1}\frac{dx_2}{dy_1} + \frac{dF}{dx_2}\frac{dx_2}{dy_2} + \frac{dF}{dy_2} &= 0. \end{aligned} \quad (5)$$

We are going to proceed by successive approximations; as a first approximation of the equations for the asymptotic surfaces, we will take Eq. (4) stopping at the second term of the series (meaning the term in $\sqrt{\mu}$ inclusive. The resulting error will then be of the same order of magnitude as μ.

1 Description of the Problem

In a second approximation, we will again take Eq. (4) for the equations of the asymptotic surfaces, but will include a larger number of terms in the series. We will be able to take a sufficiently large number of them that the resulting error will be of the same order of magnitude as μ^p, however large p might be.

Finally in the third approximation, we will try to show the properties of the *exact* equations of the asymptotic surfaces, meaning those of Eq. (2).

We therefore first need to try to form the series s_1 and s_2 from Eq. (4) directly. When substituted in place of x_1 and x_2, the series must formally satisfy Eq. (5).

We are therefore led to look for series in powers of $\sqrt{\mu}$ which formally satisfy Eq. (5). The coefficients of the series will be functions of y_1 and y_2 which will not change when y_1 and y_2 increase, respectively, by $n_1 T$ and $n_2 T$.

But we will find infinitely many series which satisfy these conditions. How can we distinguish among them those which need to go into the equalities (4)? We have seen above that in our geometric representation mode, the periodic solution being considered is represented by a closed curve and that two asymptotic surfaces pass by this closed curve. One switches from one to the other by changing $\sqrt{\mu}$ to $-\sqrt{\mu}$.

If in Eq. (2), $\sqrt{\mu}$ is therefore changed to $-\sqrt{\mu}$, then a second asymptotic surface results which must cross the first.

In other words, if the two resulting asymptotic surfaces are considered as two sheets of a single surface, it can be stated that this surface has a double curve.

Let s_1^p and s_2^p be the sum of the first p terms of the series s_1 and s_2, then the equations

$$
\begin{aligned}
x_1 &= s_1^p(y_1, y_2, \sqrt{\mu}), & x_2 &= s_2^p(y_1, y_2, \sqrt{\mu}), \\
x_1 &= s_1^p(y_1, y_2, -\sqrt{\mu}), & x_2 &= s_2^p(y_1, y_2, -\sqrt{\mu})
\end{aligned}
\tag{6}
$$

will represent two surfaces which will define the two sheets I just spoke about and which consequently must cross each other.

If these two surfaces are considered as two sheets of a single surface, it can be stated that this single surface has a double curve.

In the following we will see that this condition suffices to distinguish the series s_1 and s_2 among all the series with the same form which formally satisfy Eq. (5).

2 First Approximation

Let us continue with our usual assumptions, specifically: there are four variables, two of which are linear x_1 and x_2 and two angular y_1 and y_2, that are related by the equations

$$
\begin{aligned}
\frac{dx_1}{dt} &= \frac{dF}{dy_1}, & \frac{dx_2}{dt} &= \frac{dF}{dy_2}, \\
\frac{dy_1}{dt} &= -\frac{dF}{dx_1}, & \frac{dy_2}{dt} &= -\frac{dF}{dx_2},
\end{aligned}
\tag{1}
$$

Also, since the total energy constant C is regarded as one of the givens of the problem, these four variables satisfy the equation:

$$F(x_1, x_2, y_1, y_2) = C, \qquad (2)$$

such that only three of them are independent.

Also, a mode of geometric representation is adopted such that any state of the system corresponds to a representative point and inversely.

Also, F depends on a very small parameter μ such that F is expandable in powers of μ and written:

$$F = F_0 + \mu F_1 + \mu^2 F_2 + \cdots.$$

Also F_0 only depends on x_1 and x_2 and is independent of y_1 and y_2.

These conditions are satisfied in the specific case of the three-body problem which we used as an example in the previous section.

Let us assume that for some values of x_1 and x_2, for example for:

$$x_1 = x_1^0, \quad x_2 = x_2^0$$

the two numbers:

$$-\frac{dF_0}{dx_1} \quad \text{and} \quad -\frac{dF_0}{dx_2}$$

(that I will call for brevity n_1 and n_2) are mutually commensurable.

From what we saw in Sect. 3 of Chap. 3, for each commensurable value of the ratio n_1/n_2, there corresponds one equation

$$\frac{d\psi}{d\varpi_2} = 0,$$

(this is Eq. 7 from the section cited) and for each root of this Eq. (7) there corresponds a periodic solution of Eq. (1).

We then saw in Sect. 4 of Chap. 3 that the number of roots of Eq. (7) is always even, that half of these roots correspond to stable periodic solutions and the other half to unstable solutions.

Therefore, if μ is small enough, Eq. (1) have unstable periodic solutions.

Each of these periodic solutions will be represented by a closed trajectory curve in the adopted representation mode.

We saw in Sect. 5 of Chap. 3 that through each of these closed curves representing an unstable periodic solution, there pass two trajectory surfaces called *asymptotic*, on which are traced infinitely many trajectories which go asymptotically closer to the closed trajectory curve.

2 First Approximation

Equation (1) therefore lead us to infinitely many asymptotic trajectory surfaces for which I propose to find the equation.

First let us look at the general form of the equation for a trajectory surface. This equation can be written

$$x_1 = \Phi_1(y_1, y_2), \quad x_2 = \Phi_2(y_1, y_2),$$

where Φ_1 and Φ_2 are functions of y_1 and y_2 which must be chosen such that the following holds identically:

$$F(\Phi_1, \Phi_2, y_1, y_2) = C.$$

These two functions Φ_1 and Φ_2 will additionally have to satisfy two partial differential equations:

$$\frac{dF}{dx_1}\frac{dx_1}{dy_1} + \frac{dF}{dx_2}\frac{dx_1}{dy_2} + \frac{dF}{dy_1} = 0,$$
$$\frac{dF}{dx_1}\frac{dx_2}{dy_1} + \frac{dF}{dx_2}\frac{dx_2}{dy_2} + \frac{dF}{dy_2} = 0. \quad (3)$$

It could additionally be sufficient to consider the first of these equations, because x_2 could be eliminated from it, by replacing this variable by its value that could be determined from (2) as a function of x_1, y_1 and y_2.

We are going to proceed with integrating Eq. (3) by assuming that x_1 and x_2 are very close to x_1^0 and x_2^0 and that the ratio n_1/n_2 is commensurable.

We will now assume that x_1 and x_2 are expanded in powers of $\sqrt{\mu}$ and we will write:

$$x_1 = x_1^0 + x_1^1\sqrt{\mu} + x_1^2\mu + x_1^3\mu\sqrt{\mu} + \cdots,$$
$$x_2 = x_2^0 + x_2^1\sqrt{\mu} + x_2^2\mu + x_2^3\mu\sqrt{\mu} + \cdots, \quad (4)$$

and we will try to determine the functions x_i^k such that, by substituting their values (4) into Eq. (3) in place of x_1 and x_2, these equations can be formally satisfied.[1]

If we substitute their values (4) in place of x_1 and x_2 in F, then F will become expandable in powers of $\sqrt{\mu}$ and it will be possible to write

$$F = H_0 + \sqrt{\mu}H_1 + \mu H_2 + \mu\sqrt{\mu}H_3 + \cdots.$$

---

[1] If x_1^0 and x_2^0 were selected such that the ratio n_1/n_2 is commensurable, then one could be satisfied with expanding x_1 and x_2 in powers of μ (and not of $\sqrt{\mu}$). One would then arrive at series, which in truth would not be convergent in the geometric meaning of the word, but which like those of Mr. Lindstedt could be useful in some cases.

Furthermore it can be easily seen that[2]:

$$H_0 = F_0(x_1^0, x_2^0),$$

$$H_1 = x_1^1 \frac{dF_0}{dx_1^0} + x_2^1 \frac{dF_0}{dx_2^0} = -n_1 x_1^1 - n_2 x_2^1,$$

$$H_2 = F_1(x_1^0, x_2^0, y_1, y_2) + \left[(x_1^1)^2 \frac{d^2 F_0}{(dx_1^0)^2} + 2 x_1^1 x_2^1 \frac{d^2 F_0}{dx_1^0 dx_2^0} + (x_2^1)^2 \frac{d^2 F_0}{(dx_2^0)^2} \right] - n_1 x_1^2 - n_2 x_2^2,$$

and more generally:

$$H_k = \Theta_k - \left[L x_1^1 x_1^{k-1} + M \left(x_1^1 x_2^{k-1} + x_2^1 x_1^{k-1} \right) + N x_2^1 x_2^{k-1} \right] - n_1 x_1^k - n_2 x_2^k,$$

where Θ_k only depends on $y_1, y_2, x_1^0, x_1^1, \cdots x_1^{k-2}$ and $x_2^0, x_2^1, \cdots x_2^{k-2}$, and by setting for brevity:

$$L = -\frac{d^2 F_0}{(dx_1^0)^2}, \quad M = -\frac{d^2 F_0}{dx_1^0 dx_2^0}, \quad N = -\frac{d^2 F_0}{(dx_2^0)^2}.$$

The first of Eq. (3) then gives us, by equating equal powers of $\sqrt{\mu}$, a series of equations with which we will be able to successively determine $x_1^0, x_1^1, x_1^2, \cdots x_1^k$.

We can always assume that $n_2 = 0$. Because if that weren't the case, we could set:

$$x_1'' = a x_1 + b x_2, \quad y_1'' = d y_1 - c y_2,$$
$$x_2'' = c x_1 + d x_2, \quad y_2'' = -b y_1 + a y_2,$$

where a, b, c and d are four integers such that

$$ad - bc = 1.$$

After this change of variables, the equations still have their canonical form.

The function F which is periodic with period 2π in y_1 and y_2 is again periodic with period 2π in y_1'' and y_2''. The change of variables has therefore not altered the form of Eq. (1).

The numbers n_1 and n_2 are replaced by two new numbers n_1'' and n_2'' which relative to the transformed equations play the same role as n_1 and n_2 relative to the original equations and we have:

$$n_1'' = d n_1 - c n_2,$$
$$n_2'' = -b n_1 + a n_2.$$

---

[2][Translator: See erratum at end.]

2 First Approximation

But, since the ratio of n_1 and n_2 is commensurable by assumption, it is always possible to choose four integers a, b, c and d such that

$$n_2'' = -bn_1 + an_2 = 0.$$

We can therefore without loss of generality assume that n_2 is zero; this is what we will do until new order.

At the same time we will assume $n_1 T = 2\pi$.

If, after this simplification, we equate the coefficients of $\sqrt{\mu}$ on both sides of both Eq. (3), it will follow that

$$-n_1 \frac{dx_1^1}{dy_1} = 0 = -n_1 \frac{dx_2^1}{dy_1} = 0 \qquad (5)$$

which shows that x_1^1 and x_2^1 only depend on y_2.

We now equate the coefficients of μ on both sides of the first of Eq. (3), it then follows, making use of Eq. (5), that:

$$-n_1 \frac{dx_1^2}{dy_1} - \left(Mx_1^1 + Nx_2^1\right)\frac{dx_1^1}{dy_2} + \frac{dF_1}{dy_1} = 0. \qquad (6)$$

We propose in what follows to determine the functions x_i^k such that they are periodic functions of y_1, which must be unchanged when y_1 increases by 2π and y_2 keeps the same value.

Our functions can then be expanded in trigonometric series by sines and cosines of multiples of y_1. We will agree on the notation:

$$[U]$$

to represent the constant term in the expansion of the periodic function U along trigonometric lines of y_1 and its multiples. Under these conditions we will have:

$$\left[\frac{dU}{dy_1}\right] = 0,$$

and I can write

$$\left[\left(Mx_1^1 + Nx_2^1\right)\frac{dx_1^1}{dy_2}\right] = 0,$$

$$\left[\left(Mx_1^1 + Nx_2^1\right)\frac{dx_2^1}{dy_2}\right] = \left[\frac{dF_1}{dy_2}\right].$$

Since x_1^1 and x_2^1 do not depend on y_1, I can write more simply:

$$\frac{dx_1^1}{dy_2} = 0, \quad (Mx_1^1 + Nx_2^1)\frac{dx_2^1}{dy_2} = \left[\frac{dF_1}{dy_2}\right]. \tag{7}$$

The first of these equations shows that x_1^1 reduces to a constant. The second equation can be easily integrated. In fact it holds that:

$$\left[\frac{dF_1}{dy_2}\right] = \frac{d[F_1]}{dy_2},$$

which gives us the following for the integral of Eq. (7)

$$Mx_1^1 x_2^1 + \frac{N}{2}(x_2^1)^2 = [F_1] + C_1, \tag{8}$$

where C_1 designates a constant of integration.

But if we look at the total energy constant C as a given of the problem, we may not consider the two constants x_1^1 and C_1 is arbitrary. It must in fact hold identically that:

$$F = H_0 + \sqrt{\mu}H_1 + \mu H_2 + \mu\sqrt{\mu}H_3 + \cdots = C$$

or

$$H_0 = C, \quad H_1 = 0, \quad H_2 = 0, \quad \cdots$$

or

$$F_0(x_1^0, x_2^0) = C, \quad -n_1 x_1^1 = 0, \quad \cdots.$$

Thus the constant x_1^1 is zero which leads to additional simplifications in our equations.

In fact, Eq. (8) becomes:

$$x_2^1 = \sqrt{\frac{2}{N}(|F_1| + C_1)}.$$

In this section we will satisfy ourselves with writing and discussing the equations for our trajectory surfaces while neglecting the terms in μ and considering only terms in $\sqrt{\mu}$.

We will therefore assume that x_1 and x_2 are defined as a function of y_1 and y_2 by the following equations:

2 First Approximation

$$x_1 = x_1^0 + \sqrt{\mu} x_1^1 = x_1^0,$$

$$x_2 = x_2^0 + \sqrt{\mu} x_2^1 = x_2^0 + \sqrt{\frac{2\mu}{N}([F_1] + C_1)}.$$

Based on them, x_1 is a constant and x_2 is a only function of y_2, independent of y_1.

Let us go back to our first example from Sect. 1 of Chap. 4. What we will say would apply as well to the two other examples, but I want to focus on the first one because it is a specific case of the three-body problem.

We have seen that the state of the system could be represented by the point P which has the following rectangular coordinates:

$$\cos y_1' e^{\xi \cos y_2'}, \quad \sin y_1' e^{\xi \cos y_2'}, \quad \xi \sin y_1',$$

where

$$y_1' = \frac{1}{2}(y_1 + y_2), \quad y_1' = \frac{1}{2}(y_1 - y_2), \quad \xi = \frac{x_2 - x_1}{x_2 + x_1} = \frac{L - G}{L + G} = \frac{-x_2'}{x_1'},$$

$$y_1 = g - t, \quad y_2 = l.$$

We further observed that the variables

$$x_1' = x_1 + x_2, \quad x_2' = x_1 - x_2$$

with y_1' and y_2' form a set of canonical variables.

We can therefore regard ξ, y_1' and y_2' as a specific coordinate system defining the position of the point P in space such that any relation between ξ, y_1' and y_2' is the equation of the surface.

But next, we have to make another change of variables.

We set:

$$x_1'' = ax_1 + dx_2, \quad y_1'' = dy_1 - cy_2,$$
$$x_2'' = cx_1 + dx_2, \quad y_2'' = -by_1 + ay_2,$$

by selecting the integers a, b, c and d so as to make the number that we called n_2'' become zero.

After this change of variables, we eliminated the accents which became unnecessary and we returned to the names x_1, x_2, y_1 and y_2 for the new independent variables x_1'', x_2'', y_1'' and y_2''.

Consequently, the variables that we have called x_1, x_2, y_1 and y_2 throughout the preceding calculation, and *for which we will now retain this name*, are not the same as those that we had designated by the same letters in the first example from Sect. 1 of Chap. 4, meaning G, L, $g - t$ and l.

It is clear that our new y_1 and our new y_2 are linear functions of

$$y_1' = \tfrac{1}{2}(g - t + l) \quad \text{and} \quad y_2' = \tfrac{1}{2}(g - t - l)$$

and that the ratio of the new x_2 to the new x_1 is a linear and fractional function of ξ.

From that, we must conclude that the position of the point P in space can be fully defined by the new y_1, new y_2 and the ratio of the new x_2 to the new x_1 such that any relationship between y_1, y_2 and x_2/x_1 is the equation of a surface.

Now this specific coordinate system is such that y_1 or y_2 can be increased by a multiple of 2π without the point P changing.

The approximate equation of our trajectory surface, neglecting terms in μ, will be:

$$\frac{x_2}{x_1} = \frac{x_2^0 + x_2^1\sqrt{\mu}}{x_1^0 + x_1^1\sqrt{\mu}} = \frac{x_2^0}{x_1^0} + \frac{\sqrt{\mu}}{x_1^0}\sqrt{\frac{2}{N}([F_1] + C_1)}. \tag{9}$$

We propose first to build the surfaces represented by this approximate Eq. (9).

First we observe that $y_1 = 0$ is the equation of some surface S and that the portion of the surface that will be useful to us is a portion of contactless surface.

In fact it is sufficient to show that:

$$\frac{dy_1}{dt} \neq 0.$$

Remark that it is obviously so, because if one sets

$$F = F_0 + \mu F_1 + \mu^2 F_2 + \cdots$$

it follows that:

$$\frac{dy_1}{dt} = n_1 - \mu \frac{dF_1}{dx_1} - \mu^2 \frac{dF_2}{dx_1} + \cdots.$$

Since the parameter of μ is very small, dy_1/dt has the same sign as n_1 and n_1 is a constant which always has the same sign.

Therefore dy_1/dt always has the same sign and cannot become zero.
Which was to be proved.

The position of a point P on the surface F will be defined by two other coordinates y_2 and x_2/x_1; this coordinate system is entirely analogous to polar coordinates, meaning that the curves:

$$x_2/x_1 = \text{const.}$$

are concentric closed curves and that the point P does not change when the other coordinate, y_2, increases by 2π.

2 First Approximation

Let us go back to the surfaces defined by Eq. (9) and study their intersections with the portion of the surface S whose equation is $y_1 = 0$.

First I remark that since $\sqrt{\mu}$ is very small, these intersections will be only slightly different from the curves $x_2/x_1 = \text{const}$.

But to study the shape of these intersection curves more completely, we need to first look at the properties of the function

$$[F_1].$$

Let us resume with the notations from Sect. 3 of Chap. 3. In that section we set:

$$F_1 = \sum A \sin(m_1 y_1 + m_2 y_2 + m_3 y_3 + h),$$

where A and h are functions of x_1^0, x_2^0, x_3^0; since here we now only have two degrees of freedom, I will simply write:

$$F_1 = \sum A \sin(m_1 y_1 + m_2 y_2 + h).$$

By next setting:

$$y_1 = n_1 t, \quad y_2 = n_2 t + \varpi_2, \quad \omega = (n_1 m_1 + n_2 m_2)t + m_2 \varpi_2 + h,$$

we found:

$$F_1 = \sum A \sin \omega.$$

I will next set:

$$\psi = \sum_S A \sin \omega,$$

where the summation indicated by the sign $\sum_S$ extends over all terms such that

$$m_1 n_1 + m_2 n_2 = 0;$$

hence

$$\omega = m_2 \varpi_2 + h.$$

In the case we are looking at, n_2 is zero; hence the condition $m_1 n_1 + m_2 n_2 = 0$ reduces to $m_1 = 0$ and we have $y_2 = \varpi_2$; it therefore follows that:

$$\psi = \sum_S A \sin(m_2 \varpi_2 + h) = \sum_S A \sin(m_2 y_2 + h).$$

According to the definition of $[F_1]$, in order to obtain this quantity it is sufficient to delete from the expression for F_1 all terms where m_1 is not zero; it therefore follows:

$$[F_1] = \sum_S A \sin(m_2 y_2 + h) = \psi.$$

Thus the function that we are calling $[F_1]$ here is the same as what we were designating by ψ in Part I.

$[F_1]$ is consequently a periodic function of y_2 and this function is finite; it must therefore pass at least through one maximum and one minimum.

To be more definite, let us assume that $[F_1]$ changes in the following way when y_2 changes from 0 to 2π.

For $y_2 = 0$, $[F_1]$ goes through a maximum equal to ϕ_1.
For $y_2 = \eta_1$, $[F_1]$ goes through a minimum equal to ϕ_2.
For $y_2 = \eta_2$, $[F_1]$ goes through a maximum equal to ϕ_3.
For $y_2 = \eta_3$, $[F_1]$ goes through a minimum equal to ϕ_4.
For $y_2 = 2\pi$, $[F_1]$ returns to the value ϕ_1.

$$\varphi_1 > \varphi_2, \varphi_3 > \varphi_4.$$

These assumptions can be represented by the following curve whose abscissa is y_2 and whose ordinate is $[F_1]$ (Fig. 1).

Having thus set the ideas, I can construct the curves:

$$y_1 = 0, \quad \frac{x_2}{x_1} = \frac{x_2^0}{x_1^0} + \frac{\sqrt{\mu}}{x_1^0} \sqrt{\frac{2}{N}([F_1] + C_1)}.$$

We will see that according to the value of the integration constant C_1, these curves will take on different shapes.

In Fig. 2, I represented the two curves $C_1 = -\varphi_4$ and $C_1 = -\varphi_2$ by a solid line; these two curves each have a double point whose coordinates are respectively

$$\frac{x_2}{x_1} = \frac{x_2^0}{x_1^0}, \quad y_2 = \eta_3$$

Fig. 1 Curve representing assumptions

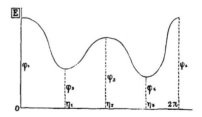

2 First Approximation

Fig. 2 Curves corresponding to different values of C_1

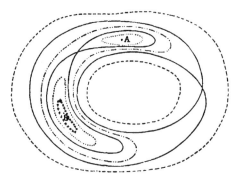

and

$$\frac{x_2}{x_1} = \frac{x_2^0}{x_1^0}, \quad y_2 = \eta_1.$$

I represented the two branches of a curve corresponding to a value of $C_1 > -\varphi_4$ by dashed line.

I represented a curve corresponding to a value of C_1 included between $-\varphi_2$ and $-\varphi_4$ by a dash-double-dot line.

I represented the two branches of a curve corresponding to a value of C_1 included between $-\varphi_2$ and $-\varphi_3$ by a dotted line.

For $C_1 = -\varphi_3$, one of these two branches reduces to a point $((x_2/x_1 = x_2^0/x_1^0), (y_2 = \eta_2))$ represented by A in the figure; the other branch is represented by the line with ×××× in the figure.

For C_1 included between $-\varphi_3$ and $-\varphi_1$, only this second branch remains; for $C_1 = -\varphi_1$, it in its turn reduces to a point represented in the figure by B and has for coordinates:

$$\frac{x_2}{x_1} = \frac{x_2^0}{x_1^0}, \quad y_2 = 0.$$

Finally for $C_1 < -\varphi_1$, the curve becomes entirely imaginary.

The surfaces defined by Eq. (1) have a general shape which can be easily deduced from those of the curves which we just constructed.

In fact, consider an arbitrary one of these curves and make one of the lines whose general equation is:

$$y_2 = \text{const.}, \quad \frac{x_2}{x_1} = \text{const.}$$

pass through each of its points.

The family of lines thus constructed will constitute a closed surface, which will be precisely one of the surfaces defined by Eq. (9).

From this it can be seen that these surfaces will in general be triply connected closed surfaces (meaning having the same connections as a torus).

For $C_1 > -\varphi_4$ or for C_1 included between $-\phi_2$ and $-\phi_3$ two similar surfaces are found, one inside the other in the first case and one outside the other in the second.

For C_1 included between $-\phi_3$ and $-\phi_1$, or between $-\phi_2$ and $-\phi_4$ there is now only a single triply connected surface; finally for $C_1 < -\varphi_1$ the surface no longer exists at all.

Now switch to four other notable surfaces:

$$C = -\varphi_1, \quad -\varphi_2, \quad -\varphi_3 \quad \text{and} \quad -\varphi_4.$$

The surfaces $C_1 = -\varphi_2$ and $C_1 = -\varphi_4$ have a double curve and have the same connections as a surface generated by the revolution of the limaçon of Pascal at a double point or a lemniscate around an axis which does not intersect the curve.

The surface $C_1 = -\varphi_3$ reduces to a single triply connected closed surface and to an isolated closed surface; finally the surface $C_1 = -\varphi_1$ reduces to an isolated closed curve.

In Sect. 3 of Chap. 3, we considered the equation:

$$\frac{d\psi}{d\varpi_2} = 0$$

which was Eq. 7 in that section; we saw that each of the roots of this equation corresponded to a periodic solution. But, in the case of interest to us, and according to a remark that we made, this equation can be written:

$$\frac{[dF_1]}{dy_2} = 0,$$

such that the periodic solutions will correspond to the maxima and minima of $[F_1]$. In the current case, these maxima, and likewise the minima, will be two in number.

We will therefore have two unstable periodic solutions corresponding to the two double curves of the surfaces $C_1 = -\varphi_2$ and $-\varphi_4$ and two stable periodic solutions, corresponding to the two isolated closed curves of the surfaces $C_1 = -\varphi_3$ and $-\varphi_1$.

Among the surfaces which are the ones that are only slightly different from the asymptotic solutions and represent them in a first approximation? Based on what we saw in Sect. 1, that would be those among them which have a double curve, meaning the surfaces $C_1 = -\varphi_4$ and $C_1 = -\varphi_2$.

3 Second Approximation

Let us return to Eq. (1) from the previous section and the assumptions made at the beginning of that section; let us write:

3 Second Approximation

$$x_1 = x_1^0 + x_1^1\sqrt{\mu} + x_1^2\mu + x_1^3\mu\sqrt{\mu} + \cdots,$$

$$x_2 = x_2^0 + x_2^1\sqrt{\mu} + x_2^2\mu + x_2^3\mu\sqrt{\mu} + \cdots;$$

now imagine that the coefficients of these two expansions are functions of y_1 and y_2 and seek to determine these coefficients such that these equations are compatible with the differential equations Eq. (1) of the previous section, meaning that it holds that:

$$\frac{dF}{dx_1}\frac{dx_1}{dy_1} + \frac{dF}{dx_2}\frac{dx_1}{dy_2} + \frac{dF}{dy_1} = 0,$$
$$\frac{dF}{dx_1}\frac{dx_2}{dy_1} + \frac{dF}{dx_2}\frac{dx_2}{dy_2} + \frac{dF}{dy_2} = 0. \tag{1}$$

That is the problem that we set for ourselves above.

This problem can be placed in another form (by taking the perspective of *Vorlesungen über Dynamik* [C.G.J. Jacobi, 1884]).

If x_1 and x_2 are two functions of y_1 and y_2 which satisfy Eq. (1), then the expression:

$$x_1 dy_1 + x_2 dy_2$$

will have to be an exact differential. If we therefore set:

$$dS = x_1 dy_1 + x_2 dy_2,$$

S will be a function of y_1 and y_2 which will be defined by the partial differential equation:

$$F\left(\frac{dS}{dy_1}, \frac{dS}{dy_2}, y_1, y_2\right) = C. \tag{2}$$

We can expand S in powers of $\sqrt{\mu}$ and we will obtain:

$$S = S_0 + S_1\sqrt{\mu} + S_2\mu + S_3\mu\sqrt{\mu} + \cdots. \tag{3}$$

$S_0, S_1, \ldots S_K, \ldots$ will be functions of y_1 and y_2 and we will have:

$$\frac{dS_k}{dy_1} = x_1^k, \quad \frac{dS_k}{dy_2} = x_2^k.$$

I now review some conditions that we imposed in the previous section on the functions x_1^k and x_2^k; we first assumed that x_1^0 and x_2^0 would need to be constants. Then one has:

$$S_0 = x_1^0 y_1 + x_2^0 y_2.$$

If we next call n_1 and n_2 the values of $-dF_0/dx_1$ and $-dF_0/dx_2$ for $x_1 = x_1^0$, $x_2 = x_2^0$; these quantities n_1 and n_2 will again be constants. The analysis which is going to follow applies to the case where the ratio n_1/n_2 is commensurable. In this case it is always possible, as we have seen, to assume $n_2 = 0$; and that is what we are going to do now just as we did in the previous section.

We furthermore assumed in that section that x_1^k and x_2^k are periodic functions of y_1 which do not change value when y_1 and y_2 change to $y_1 + 2\pi$ and y_2.

From this it follows that dS_k/dy_1 and dS_k/dy_2 are periodic functions with respect to y_1 and that we can write:

$$S_k = \frac{\lambda_k}{n_1} y_1 + S'_k, \tag{4}$$

where λ_k is a constant and S'_k is a periodic function of y_1.

Let us assume that in the left-hand side of Eq. (2)

$$F\left(\frac{dS}{dy_1}, \frac{dS}{dy_2}, y_1, y_2\right)$$

we replace S by its expansion (3); it will then be seen that F is expandable in powers of $\sqrt{\mu}$ and then, as was seen in the previous section, it will follow that:

$$F = H_0 + H_1\sqrt{\mu} + H_2\mu + H_3\mu\sqrt{\mu} + \cdots,$$

where the H are functions of y_1, y_2 and partial derivatives of S_0, S_1, S_2, etc.

It can additionally be seen that H_0 will depend only on S_0; H_1 on S_0 and S_1; H_2 on S_0, S_1 and S_2; H_1 on S_0, S_1, S_2 and S_3; etc.

It is additionally found that:

$$H_0 = F_0(x_1^0, x_2^0) = C,$$
$$H_1 = -n_1 \frac{dS_1}{dy_1},$$
$$H_2 = -n_1 \frac{dS_2}{dy_1} + \Delta S_1 + K_2,$$
$$H_3 = -n_1 \frac{dS_3}{dy_1} + 2\Delta S_2 + K_3,$$
$$\vdots$$
$$H_p = -n_1 \frac{dS_p}{dy_1} + 2\Delta S_{p-1} + K_p,$$

3 Second Approximation

where for brevity we set:

$$\Delta S_p = \frac{1}{2}\left[\frac{d^2 F_0}{(dx_1^0)^2} x_1^1 x_1^p + \frac{d^2 F_0}{dx_1^0 dx_2^0}\left(x_1^1 x_2^p + x_2^1 x_1^p\right) + \frac{d^2 F_0}{(dx_2^0)^2} x_2^1 x_2^p\right]$$

and where K_p depends only on $S_0, S_1, \ldots$ up to S_{p-2}.

With that set and in order to determine the functions S_p recursively, we will have the following equations:

$$H_0 = C, \quad H_1 = 0, \quad H_2 = 0, \quad \cdots \quad H_p = 0.$$

If it were assumed that the functions $S_0, S_1, \ldots S_{p-1}$ had been fully known, then the equation

$$H_p = 0$$

or

$$n_1 \frac{dS_p}{dy_1} = 2\Delta S_{p-1} + K_p \tag{5}$$

would determine the function S_p within an arbitrary function of y_2.

But that isn't entirely how the question is presented.

Assume that

$$S_0, S_1, \ldots S_{p-2}$$

are fully known and that S_{p-1} is known within an arbitrary function of y_2.

By assumption, the derivatives of $S_0, S_1, \ldots S_{p-2}, S_{p-1}$ are periodic functions of y_1; therefore K_p and ΔS_{p-1} will be periodic functions of y_1.

As we did in the previous section, designate the average value of U, which is a periodic function of y_1, by $[U]$.

S_p must be in the form (4); from that we conclude that:

$$\left[\frac{dS_p}{dy_1}\right]$$

must be a constant λ_p/n_1 independent of y_2, such that Eq. (5) gives us:

$$2[\Delta S_{p-1}] + [K_p] = \lambda_p, \tag{6}$$

and this equation will fully determine S_{p-1} (if it is assumed that the constant λ_p is given either arbitrarily or by some law).

First we find the equation:

$$H_1 = 0 \quad \text{or} \quad \frac{dS_1^0}{dy_1} = 0$$

which shows us that S_1 is an arbitrary function of y_2.

From this we will deduce:

$$2\Delta S_p = -M \frac{dS_1}{dy_2} \frac{dS_p}{dy_1} - N \frac{dS_1}{dy_2} \frac{dS_p}{dy_2}$$

(for brevity we set:

$$-M = \frac{d^2 F_0}{dx_1^0 dx_2^0}, \quad -N = \frac{d^2 F_0}{(dx_2^0)^2}$$

as we did in the section cited).

The following is the equation that we next find by setting the average value of H_2 equal to 0:

$$2[\Delta S_1] + [K_2] = \lambda_1.$$

Hence

$$\Delta S_1 = -\frac{N}{2} \left(\frac{dS_1}{dy_2} \right)^2 = [\Delta S_1].$$

Furthermore:

$$K_2 = F_1\left(x_1^0, x_2^0, y_1, y_2\right).$$

λ_2 is a constant which, as can be easily seen, is precisely what we called $-C_1$ in the section cited.

It therefore follows:

$$\frac{dS_1}{dy_2} = \sqrt{\frac{2}{N}([F_1] + C_1)}.$$

S_1 is in that way completely determined up to a constant; but we can set this constant aside, it in fact does not play any role because the functions S only enter through their derivatives.

3 Second Approximation

Equation (6) then becomes:

$$\left[N \frac{dS_1}{dy_2} \frac{dS_{p-1}}{dy_2}\right] = -\lambda_p - M\left[\frac{dS_1}{dy_2} \frac{dS_{p-1}}{dy_1}\right] + [K_p]. \tag{7}$$

Everything on the right-hand side is known; K_p depends only on $S_0, S_1, \ldots S_{p-2}$; dS_{p-1}/dy_1 is known because S_{p-1} is assumed to be determined up to an arbitrary function of y_2.

Additionally dS_1/dy_2 is independent of y_1; the left-hand side can therefore be written:

$$N \frac{dS_1}{dy_1}\left[\frac{dS_{p-1}}{dy_2}\right],$$

such that Eq. (7) will give us $[dS_{p-1}/dy_2]$ as a function of y_2. We will therefore know $[S_{p-1}]$ up to a constant and this constant, which plays no role, can be put aside.

We know both S_{p-1} to within an arbitrary function of y_2 and also $[S_{p-1}]$ as a function of y_2; therefore S_{p-1} is fully determined.

The constant C_1 plays a dominant role. Let us first assume that it is greater than the value which we've called $-\varphi_4$ in the sections cited and that consequently $[F_1] + C_1$ is always positive and dS_1/dy_2 is always real and I can add that it is always positive because I am free to choose the $+$ in front of the radical.

I state that in this case, the constants λ can be chosen arbitrarily and that dS_p/dy_1 and dS_p/dy_2 are periodic functions not only of y_1, but also of y_2 (S_p is then of the form

$$S_p = \lambda_p y_1 + \mu_p y_2 + S_p'',$$

where λ_p and μ_p are constants while S_p'' is periodic with period 2π both with respect to y_1 and with respect to y_2.).

In fact, let us assume that this is true for:

$$\frac{dS_1}{dy_1}, \frac{dS_1}{dy_2}, \frac{dS_2}{dy_1}, \frac{dS_2}{dy_2}, \ldots \frac{dS_{p-2}}{dy_1}, \frac{dS_{p-2}}{dy_2}, \frac{dS_{p-1}}{dy_1};$$

I state that this will also be true for dS_{p-1}/dy_2 and dS_p/dy_1.
In fact, by assumption we have:

$$\frac{dS_{p-1}}{dy_1} = \sum A_{m_1 m_2} \cos(m_1 y_1 + m_2 y_2 + \alpha),$$

where the A and the α are constants, and m_1 and m_2 are integers.

Next by definition it follows that:

$$\left[\frac{dS_{p-1}}{dy_1}\right] = \sum_{m_2=0}^{\infty} A_{0,m_2} \cos(m_2 y_2 + \alpha).$$

But it must hold that

$$2[\Delta S_{p-2}] + [K_{p-1}] = \lambda_{p-1}$$

and consequently:

$$\left[\frac{dS_{p-1}}{dy_1}\right] = \frac{\lambda_{p-1}}{n_1},$$

where λ_{p-1} is a constant; from this it is concluded that:

$$A_{0,m_2} = 0 \text{ for } m_2 \neq 0, \quad A_{0,0} = \frac{\lambda_{p-1}}{n_1}.$$

In that way it follows that

$$S_{p-1} = \frac{\lambda_{p-1}}{n_1} y_1 + \sum A_{m_1 m_2} \frac{\sin(m_1 y_1 + m_2 y_2 + \alpha)}{m_1} + [S_{p-1}],$$

where m_1 and m_2 under the summation sign always take all integer values such that $m_1 \neq 0$.

Thus, in order for S_{p-1} to have the desired form, it is sufficient that

$$\left[\frac{dS_{p-1}}{dy_2}\right]$$

be a periodic function of y_2. Hence $[dS_{p-1}/dy_2]$ is defined by the equation:

$$N \frac{dS_1}{dy_2} \left[\frac{dS_{p-1}}{dy_2}\right] = -\lambda_p - M \frac{dS_1}{dy_2} \left[\frac{dS_{p-1}}{dy_1}\right] + [K_p].$$

where K_p, which depends only on $S_1, S_2, \ldots S_{p-2}$, will be periodic in y_2.

$[dS_{p-1}/dy_1]$ is a constant λ_{p-1}/n_1; additionally, dS_1/dy_2 is a periodic function of y_2 *which never becomes zero.*

It follows from this that $[dS_{p-1}/dy_2]$ is expandable in sines and cosines of multiples of y_2.

3 Second Approximation

It next follows that:

$$n_1 \frac{dS_p}{dy_1} = 2(\Delta S_{p-1}) + K_p,$$

which shows that dS_p/dy_1 is periodic in y_1 and y_2.

In that way by choosing a value for C_1 greater than $-\varphi_4$ and by next arbitrarily choosing the other constants $\lambda_3, \lambda_4, \cdots$, ordered series of sines and cosines of multiples of y_1 and y_2 are found for dS/dy_1 and dS/dy_2. Even though divergent, the series can be useful in certain cases.

Now take up the case where

$$C_1 = -\varphi_4,$$

which as we saw in Sect. 2 is the one which corresponds to the series which the asymptotic surfaces represent asymptotically.

The expression

$$[F_1] + C_1$$

is never negative, but it becomes zero for some value of y_2 which we have called η_3 in the referenced section. In this section I will assume that this value is zero; I have the right to do that because the choice only implies a specific selection of the origin of y_2.

Therefore we write $[F_1] + C_1$ in trigonometric series form:

$$[F_1] + C_1 = \sum A_m \sin my_2 + \sum B_m \cos my_2.$$

For $y_2 = 0$, this function and also its derivative become zero; because the function is always positive, zero is a minimum of the function. From this it results that the following expression:

$$\frac{[F_1] + C_1}{\sin^2 y_2/2}$$

are expandable in sines and cosines of multiples of y_2; it is a periodic function of y_2 which never becomes zero and never becomes infinite.

It follows from that that it is possible to write:

$$\frac{\sin y_2/2}{\sqrt{[F_1] + C_1}} = \sum A_m \cos my_2 + \sum B_m \sin my_2$$

and consequently:

$$\frac{dS_1}{dy_2} = \frac{\sqrt{2/N}\sin y_2/2}{\sum A_m \cos my_2 + \sum B_m \sin my_2}.$$

We can now write Eq. (7) in the following form:

$$\frac{\sqrt{2N}\sin y_2/2}{\sum A_m \cos my_2 + \sum B_m \sin my_2}\left[\frac{dS_{p-1}}{dy_2}\right] = -\lambda_p + \Phi_p(y_2), \qquad (7')$$

where Φ_p is a known function of y_2.

Having laid that out, I propose to prove that:

$$\frac{dS_p}{dy_1} \quad \text{and} \quad \frac{dS_p}{dy_2}$$

are periodic functions of y_1 and y_2, whose period is 2π in y_1 and 4π in y_2.

Let us assume in fact that this was demonstrated for:

$$\frac{dS_1}{dy_1}, \frac{dS_1}{dy_2}, \frac{dS_2}{dy_1}, \frac{dS_2}{dy_2}, \ldots \frac{dS_{p-2}}{dy_1}, \frac{dS_{p-2}}{dy_2}, \frac{dS_{p-1}}{dy_1}.$$

dS_{p-1}/dy_1 is a periodic function of y_1 and y_2; furthermore, its average value

$$\left[\frac{dS_{p-1}}{dy_1}\right] = \frac{1}{n_1}\lambda_{p-1}$$

is a constant independent of y_2. We can therefore write:

$$S_{p-1} = \frac{1}{n_1}\lambda_{p-1}y_1 + \theta_{p-1}(y_1,y_2) + \zeta_{p-1}(y_2).$$

where $\theta_{p-1}(y_1,y_2)$ is a periodic function of y_1 and y_2 and ζ_{p-1} is an arbitrary function of only y_2. It next follows:

$$\frac{dS_{p-1}}{dy_2} = \frac{d\theta_{p-1}}{dy_2} + \frac{d\zeta_{p-1}}{dy_2},$$

hence

$$\frac{d[S_{p-1}]}{dy_2} = \frac{d[\theta_{p-1}]}{dy_2} + \frac{d[\zeta_{p-1}]}{dy_2}$$

3 Second Approximation

and

$$\frac{dS_{p-1}}{dy_2} - \frac{d[S_{p-1}]}{dy_2} = \frac{d\theta_{p-1}}{dy_2} - \frac{d[\theta_{p-1}]}{dy_2},$$

which shows that $dS_{p-1}/dy_2 - d[S_{p-1}]/dy_2$ is a periodic function of y_1 and y_2.

Equation (7′) shows that this is also true for $[dS_{p-1}/dy_2]$ and consequently for dS_{p-1}/dy_2 (whatever additionally the constant λ_p might be) and Eq. (5) shows that this is true for dS_p/dy_1.

This will therefore be true for the functions:

$$\frac{dS_p}{dy_1} \quad \text{and} \quad \frac{dS_p}{dy_2}$$

whatever the index p.

It is all the same important to remark that while these functions are periodic, this is not a sufficient reason for them to be expandable in sines and cosines of multiples of y_1 and $y_2/2$. In fact these functions are not always finite, except for a specific choice of the constant λ_p; it is easy to see that this is so because Eq. (7′) from which the value

$$\left[\frac{dS_{p-1}}{dy_2}\right]$$

must be taken has a factor of $\sin y_2/2$ in its left-hand side. Therefore the expression for $[dS_{p-1}/dy_2]$ will contain $\sin y_2/2$ in the denominator.

The derivatives of the functions S_p will therefore become infinite, but only for

$$\sin\frac{y_2}{2} = 0 \quad \text{or} \quad y_2 = 2k\pi.$$

If y_2 has a value other than $2k\pi$, these derivatives will not become infinite for any value of y_1; they can therefore be expanded in sines and cosines of multiples of y_1.

We can therefore write for example:

$$\frac{dS_{p-1}}{dy_1} = \frac{1}{n_1}\lambda_{p-1} = \sum A_m \cos my_1 + \sum B_m \sin my_1$$

where A_m and B_m are periodic functions of y_2 which can become infinite.

Now imagine that the constants λ_p with odd index are all zero; I state that

$$\frac{dS_p}{dy_1} \quad \text{and} \quad \frac{dS_p}{dy_2}$$

will not change when y_2 is changed to $y_2 + 2\pi$ whenever the index p is even and on the other hand that these two functions will change sign, without changing absolute value, when y_2 is changed to $y_2 + 2\pi$ whenever the index p is odd.

I assume this theorem will also be true for:

$$\frac{dS_1}{dy_1}, \frac{dS_1}{dy_2}, \frac{dS_2}{dy_1}, \frac{dS_2}{dy_2}, \ldots \frac{dS_{p-2}}{dy_1}, \frac{dS_{p-2}}{dy_2}, \frac{dS_{p-1}}{dy_1}$$

and I propose to prove that it is also true for

$$\frac{dS_{p-1}}{dy_2} \quad \text{and} \quad \frac{dS_p}{dy_1}.$$

If dS_{p-1}/dy_1 is multiplied by $(-1)^{p-1}$ when y_2 is changed to $y_2 + 2\pi$, then it will be the same for:

$$\frac{dS_{p-1}}{dy_2} - \left[\frac{dS_{p-1}}{dy_2}\right].$$

We in fact found

$$\frac{dS_{p-1}}{dy_1} = \frac{1}{n_1}\lambda_{p-1} = \sum A_m \cos my_1 + \sum B_m \sin my_1,$$

where A_m and B_m are periodic functions of y_2.

If dS_{p-1}/dy_1 is multiplied by $(-1)^{p-1}$ when y_2 is increases by 2π, then it will be the same for A_m and B_m and for derivatives of these functions with respect to y_2. It will therefore still be the same for:

$$\frac{dS_{p-1}}{dy_2} - \left[\frac{dS_{p-1}}{dy_2}\right] = \sum \frac{dA_m}{dy_2}\frac{\sin my_1}{m} - \sum \frac{dB_m}{dy_2}\frac{\cos my_1}{m}.$$

We now have to show that this is true for

$$\left[\frac{dS_{p-1}}{dy_2}\right].$$

In order to do that it is necessary to study how K_p depends on the functions $S_0, S_1, S_2, \ldots S_{p-1}$. I propose to establish that the order of all the terms of K_p with respect to the derivatives of functions with odd index

$$S_1, S_3, S_5, \ldots$$

will have the same parity as p.

3 Second Approximation

In fact by setting

$$S = S_0 + S_1\sqrt{\mu} + S_2\mu + \cdots$$

in

$$F\left(\frac{dS}{dy_1}, \frac{dS}{dy_2}, y_1, y_2\right),$$

we found:

$$F = H_0 + H_1\sqrt{\mu} + H_2\mu + \cdots.$$

If I change $\sqrt{\mu}$ into $-\sqrt{\mu}$ at the same time as I change S_1, S_3, S_5, etc., into $-S_1, -S_3, -S_5$ etc., without touching the functions of even index, the expression for F will not have to change.

Therefore H_p will not have to change into $(-1)^p H_p$.

That shows that the order of all the terms of H_p with respect to the derivatives of S_1, S_3, S_5, etc., will have to have the same parity as p. It will therefore have to be, as I stated, the same for the terms of K_p because K_p was obtained by deleting the terms which depend on S_{p-1} or S_p from H_p.

Having stated that, let us change y_2 to $y_2 + 2\pi$; the derivatives of S_{p-1} will not change if q is even and at most equal to $p - 2$; they will change sign if q is odd and at most equal to $p - 2$. Therefore K_p will change into $(-1)^p K_p$.

Now let us go back to Eq. (7)

$$N\frac{dS_1}{dy_2}\left[\frac{dS_{p-1}}{dy_2}\right] = -\lambda_p - M\frac{dS_1}{dy_2}\left[\frac{dS_{p-1}}{dy_1}\right] + [K_p]. \tag{7}$$

When y_2 is changed to $y_2 + 2\pi$,

$$[K_p] \text{ changes to } (-1)^p [K_p],$$
$$\left[\frac{dS_{p-1}}{dy_1}\right] \text{ changes to } (-1)^p \left[\frac{dS_{p-1}}{dy_1}\right],$$

and

$$\frac{dS_1}{dy_2} \text{ changes to } -\frac{dS_1}{dy_2}.$$

We can even state that

$$\lambda_p \text{ changes to } (-1)^p \lambda_p.$$

In fact this is true for even p because λ_p is a constant independent of y_2; this is also true for odd p because we have assumed that the λ_p with odd index are all zero.

From this it follows that

$$\left[\frac{\mathrm{d}S_{p-1}}{\mathrm{d}y_2}\right] \quad \text{changes to} \quad (-1)^{p-1}\left[\frac{\mathrm{d}S_{p-1}}{\mathrm{d}y_2}\right]$$

and consequently

$$\frac{\mathrm{d}S_{p-1}}{\mathrm{d}y_2} \quad \text{changes to} \quad (-1)^{p-1}\frac{\mathrm{d}S_{p-1}}{\mathrm{d}y_2}.$$

Which was to be proved.

I now state that $\mathrm{d}S_p/\mathrm{d}y_1$ will change to $(-1)^p \mathrm{d}S_p/\mathrm{d}y_1$.

We next write Eq. (5)

$$n_1 \frac{\mathrm{d}S_p}{\mathrm{d}y_1} = 2\Delta S_{p-1} + K_p; \tag{5}$$

K_p and ΔS_{p-1} and consequently the right-hand side of Eq. (5) will be multiplied by $(-1)^p$ when y_2 increases by 2π. It will therefore need to be the same for the left-hand side and for $\mathrm{d}S_p/\mathrm{d}y_1$.

Which was to be proved.

I am now going to prove that the constants λ_p can be chosen so that the derivatives of the functions S_p do not become infinite for $y_2 = 2k\pi$.

Let us assume that the constants $\lambda_2, \lambda_3, \ldots \lambda_{p-1}$ were chosen such that

$$\frac{\mathrm{d}S_1}{\mathrm{d}y_1}, \frac{\mathrm{d}S_1}{\mathrm{d}y_2}, \ldots \frac{\mathrm{d}S_{p-2}}{\mathrm{d}y_1}, \frac{\mathrm{d}S_{p-2}}{\mathrm{d}y_2}, \frac{\mathrm{d}S_{p-1}}{\mathrm{d}y_1}$$

remain finite and that the constants λ_q with odd index are zero; I propose to choose λ_p such that $\mathrm{d}S_{p-1}/\mathrm{d}y_2$ and $\mathrm{d}S_p/\mathrm{d}y_1$ don't become infinite either. We will at the same time see that λ_p will have to be zero if p is odd.

First, it is clear that if $\mathrm{d}S_{p-1}/\mathrm{d}y_1$ remains finite, then so will:

$$\frac{\mathrm{d}S_{p-1}}{\mathrm{d}y_2} - \left[\frac{\mathrm{d}S_{p-1}}{\mathrm{d}y_2}\right]$$

and

$$\Phi_p(y_2) = [K_p] - M\frac{\mathrm{d}S_1}{\mathrm{d}y_2}\left[\frac{\mathrm{d}S_{p-1}}{\mathrm{d}y_1}\right].$$

3 Second Approximation

Now go back to Eq. (7′). The coefficient of the unknown quantity $[dS_{p-1}/dy_2]$ becomes zero for $y_2 = 2\pi k$; for this unknown quantity to remain finite, the right-hand side also has to become zero and it must be that:

$$\Phi_p(2k\pi) = \lambda_p.$$

Since Φ_p does not change when y_2 increases by 4π, it will be sufficient to take $k = 0$ and $k = 1$ and to write

$$\Phi_p(0) = \Phi_p(2\pi) = \lambda_p. \qquad (8)$$

If p is even, there is no difficulty, one has:

$$\Phi_p(y_2) = \Phi_p(y_2 + 2\pi)$$

and consequently:

$$\Phi_p(0) = \Phi_p(2\pi),$$

such that it is sufficient to take:

$$\lambda_p = \Phi_p(0).$$

If on the other hand p is odd, then we have:

$$\Phi_p(y_2) = -\Phi_p(y_2 + 2\pi)$$

and

$$\Phi_p(0) = -\Phi_p(2\pi),$$

such that Eq. (8) can only be satisfied if it holds that:

$$\Phi_p(0) = \Phi_p(2\pi) = \lambda_p = 0.$$

We therefore have to prove that for odd p, $\Phi_p(0)$ is zero.
Let it in fact be:

$$\Phi_p(0) = \alpha$$

and consequently

$$\Phi_p(2\pi) = -\alpha.$$

I state that α is zero.
We are now going to rely on a lemma that is nearly obvious.

This is the statement of this lemma:

Let φ_1 and φ_2 be two periodic functions with period 2π with respect to y_1 and to y_2. It is known that if ϕ is a periodic function of y_1, for example, then the average value of $d\varphi/dy_1$ is zero. It will then be that:

$$\iint \frac{d\varphi_1}{dy_2} dy_1 dy_2 = \iint \frac{d\varphi_2}{dy_1} dy_1 dy_2 = 0$$

or

$$\iint \left(\frac{d\varphi_1}{dy_2} - \frac{d\varphi_2}{dy_1} \right) dy_1 dy_2 = 0,$$

where the integrals are taken over all values of y_1 and y_2 from 0 to 2π.

For the lemma to be true, it is necessary that the functions φ_1 and φ_2 be continuous, but their derivatives can be discontinuous. These derivatives only need to remain finite.

Having laid that out, we will accomplish the determination of the function S_{p-1} no longer by Eq. (7') but by the following equation:

$$\frac{\sqrt{2N} \sin \frac{y_2}{2} \left[\frac{dS_{p-1}}{dy_2} \right]}{\sum A_m \cos my_2 + \sum B_m \sin my_2} = -\alpha \cos \frac{y_2}{2} + \Phi_p(y_2). \tag{9}$$

The only difference from Eq. (7') is that λ_p has been replaced by $\alpha \cos y_2/2$.

This equation shows first that $[dS_{p-1}/dy_2]$ is a periodic function of y_2 with period 2π (recall that p is assumed to be odd). Additionally this function does not become infinite for $y_2 = 2k\pi$, because the right-hand side of Eq. (9) becomes zero for $y_2 = 0$ and for $y_2 = 2\pi$.

Let us next set

$$\zeta_1 = \frac{dS_0}{dy_1} + \mu^{\frac{1}{2}} \frac{dS_1}{dy_1} + \mu \frac{dS_2}{dy_1} + \cdots \mu^{\frac{p-1}{2}} \frac{dS_{p-1}}{dy_1} + \mu^{\frac{p}{2}} \eta,$$

$$\zeta_2 = \frac{dS_0}{dy_2} + \mu^{\frac{1}{2}} \frac{dS_1}{dy_2} + \mu \frac{dS_2}{dy_2} + \cdots \mu^{\frac{p-1}{2}} \frac{dS_{p-1}}{dy_2};$$

η will be a function of y_1, y_2 and μ defined by the equation:

$$F(\zeta_1, \zeta_2, y_1, y_2) = C. \tag{10}$$

It is easy to see that ζ_2 is entirely determined because we now fully know $S_0, S_1, \ldots S_{p-1}$. Therefore η can be drawn from Eq. (10) in the following form:

3 Second Approximation

$$\eta = \eta_0 + \mu^{\frac{1}{2}}\eta_1 + \mu\eta_2 + \cdots,$$

where the η_i are periodic functions of y_1 and y_2, with period 2π with respect to y_1 and 4π with respect to y_2.

Additionally it will hold that:

$$\frac{\mathrm{d}\zeta_1}{\mathrm{d}y_2} - \frac{\mathrm{d}\zeta_2}{\mathrm{d}y_1} = \mu^{\frac{p}{2}}\frac{\mathrm{d}\eta}{\mathrm{d}y_2}.$$

We only need η_0; it is immediately seen that η_0 is given by the following equation:

$$n_1 \eta_0 = 2\Delta S_{p-1} + K_p; \tag{11}$$

the only difference between this equation and Eq. (5) is that the unknown in it is designated by η_0.

This equation shows that η_0 is a periodic function of y_1; the average value of this function needs to be found. Referring to the meaning of Eq. (9), it will be seen that it states that the average of the right-hand side of (11) is $\alpha \cos(y_2/2)$. It therefore holds that:

$$[\eta_0] = \frac{\alpha}{n_1} \cos\frac{y_2}{2}.$$

ζ_2 could have two different values which are permutations of each other, either on changing $\sqrt{\mu}$ to $-\sqrt{\mu}$, or changing y_2 to $y_2 + 2\pi$.

I will call φ_2 the larger of these two values of ζ_2 and ψ_2 the smaller.

Similarly ζ_1 could have two values; I will call φ_1 the one which corresponds to φ_2 and ψ_1 the one which corresponds to ψ_2.

Finally, η could have two values; I will call η' the one which corresponds to φ_2 and η'' the one which corresponds to ψ_2; η_i could have two values which I will similarly call η'_i and η''_i.

The function φ_2 is periodic with period 2π with respect to y_2; in fact, when y_2 increases by 2π, the two values of ζ_2 are exchanged; therefore φ_2 which is always equal to larger these two values does not change.

For the same reason, φ_1, ψ_1, ψ_2, η', η'', η'_i and η''_i will be functions with period 2π with respect to y_2.

From the preceding definitions, it results that φ_1, φ_2, ψ_1 and ψ_2 are continuous functions, although the derivatives of these functions, and also η' and η'' can be discontinuous.

We are now in the conditions where our lemma is applicable and we can write:

$$\mu^{\frac{p}{2}}\iint \frac{d\eta'}{dy_2}dy_1 dy_2 = \iint \left(\frac{d\varphi_1}{dy_2} - \frac{d\varphi_2}{dy_1}\right) dy_1 dy_2 = 0,$$

$$\mu^{\frac{p}{2}}\iint \frac{d\eta''}{dy_2}dy_1 dy_2 = \iint \left(\frac{d\psi_1}{dy_2} - \frac{d\psi_2}{dy_1}\right) dy_1 dy_2 = 0,$$

or even

$$\iint \frac{d(\eta' - \eta'')}{dy_2} dy_1 dy_2 = 0,$$

or finally

$$\iint \frac{d(\eta'_0 - \eta''_0)}{dy_2} dy_1 dy_2 + \iint \left[\frac{d(\eta' - \eta'_0)}{dy_2} - \frac{d(\eta'' - \eta''_0)}{dy_2}\right] dy_1 dy_2 = 0.$$

This relation will have to hold whatever μ is.
But when μ approaches 0, $\eta' - \eta'_0$ and $\eta'' - \eta''_0$ approach 0.
Therefore it will hold that:

$$\lim_{\mu \to 0} \iint \frac{d(\eta'_0 - \eta''_0)}{dy_2} dy_1 dy_2 = 0. \tag{12}$$

Let us transform the left-hand side of the equality (12). I first note that since p is an odd integer, η_0 is a function which must change to $-\eta_0$ when y_2 changes to $y_2 + 2\pi$. To be convinced of this, it is sufficient to refer to Eq. (11). We therefore have:

$$\eta'_0 = -\eta''_0 = \pm\eta_0$$

hence

$$\iint \frac{d(\eta'_0 - \eta''_0)}{dy_2} dy_1 dy_2 = 2\iint \frac{d\eta'_0}{dy_2} dy_1 dy_2 = 2\iint \frac{d(\pm\eta_0)}{dy_2} dy_1 dy_2.$$

It remains to be seen for which values of the y we must make $\eta'_0 = +\eta_0$ and for which values of the y we must make $\eta'_0 = -\eta_0$.
If we have:

$$\frac{dS_1}{dy_2} + \mu\frac{dS_3}{dy_2} + \mu^2\frac{dS_5}{dy_2} + \cdots \mu^{\frac{p-3}{2}}\frac{dS_{p-2}}{dy_2} > 0, \tag{13}$$

3 Second Approximation

we will need to take, per our convention:

$$\varphi_2 = \frac{dS_0}{dy_2} + \sqrt{\mu}\frac{dS_1}{dy_2} + \mu\frac{dS_2}{dy_2} + \mu\sqrt{\mu}\frac{dS_3}{dy_2} + \cdots \mu^{\frac{p-1}{2}}\frac{dS_{p-1}}{dy_2}$$

and

$$\psi_2 = \frac{dS_0}{dy_2} - \sqrt{\mu}\frac{dS_1}{dy_2} + \mu\frac{dS_2}{dy_2} - \mu\sqrt{\mu}\frac{dS_3}{dy_2} + \cdots \mu^{\frac{p-1}{2}}\frac{dS_{p-1}}{dy_2}.$$

If on the other hand the left-hand side of the inequality (13) is negative, we will need to take:

$$\varphi_2 = \frac{dS_0}{dy_2} - \sqrt{\mu}\frac{dS_1}{dy_2} + \cdots \mu^{\frac{p-1}{2}}\frac{dS_{p-1}}{dy_2}$$

and

$$\psi_2 = \frac{dS_0}{dy_2} + \sqrt{\mu}\frac{dS_1}{dy_2} + \cdots \mu^{\frac{p-1}{2}}\frac{dS_{p-1}}{dy_2}.$$

Everything depends therefore on the sign of the left-hand side of the inequality (13). Let us set this left-hand side to 0; we will obtain and equation:

$$\frac{dS_1}{dy_2} + \mu\frac{dS_3}{dy_2} + \cdots = 0. \tag{14}$$

This equation can be regarded as defining y_2 as a function of y_1 and μ.
We could solve this equation and write:

$$y_2 = \theta(y_1, \mu).$$

Observe only that θ is a periodic function with period 2π with respect to y_1 and that this function θ becomes identically equal to zero when μ is set to 0 in it.

Consequently when y_2 varies from θ to $\theta + 2\pi$, we will have:

$$\eta'_0 = +\eta_0$$

and when y_2 varies from $\theta + 2\pi$ to $\theta + 4\pi$ we will have:

$$\eta'_0 = -\eta_0.$$

Our integrals must be extended to all values of y_2 included between 0 and 2π. But, since η'_0 is a function with period 2π, we will have:

$$\int_0^{2\pi}\left[\int_0^{2\pi}\frac{d\eta'_0}{dy_2}dy_2\right]dy_1 = \int_0^{2\pi}\left[\int_\theta^{\theta+2\pi}\frac{d\eta'_0}{dy_2}dy_2\right]dy_1$$

or

$$\iint\frac{d\eta'_0}{dy_2}dy_1 dy_2 = \int_0^{2\pi}\left[\int_\theta^{\theta+2\pi}\frac{d\eta_0}{dy_2}dy_2\right]dy_1.$$

When μ approaches 0, the left-hand side will approach 0 and additionally θ will approach 0; we will therefore have:

$$\lim_{\mu\to 0}\iint\frac{d\eta'_0}{dy_2}dy_1 dy_2 = \int_0^{2\pi}\left[\int_0^{2\pi}\frac{d\eta_0}{dy_2}dy_2\right]dy_1 = 0$$

hence

$$0 = \iint\frac{d\eta_0}{dy_2}dy_1 dy_2 = 2\pi\int_0^{2\pi}\frac{d[\eta_0]}{dy_2}dy_2 = -\pi\frac{\alpha}{n_1}\int_0^{2\pi}\sin\frac{y_2}{2}dy_2 = -\frac{4\pi\alpha}{n_1}$$

We therefore have

$$\alpha = 0.$$

Which was to be proved.

The result of this is that if the constants λ_p become zero for odd index and if suitable values are given to the constants λ_p with even index, then the functions dS_p/dy_1 and dS_p/dy_2 will remain finite.

One could then expand them according to sines and cosines of multiples of y_1 and $y_2/2$; only the even multiples of $y_2/2$ will enter into the expansion if p is even; if on the other hand p is odd, only the odd multiples of $y_2/2$ will enter.

For the approximate equations for the asymptotic surfaces we will then have:

$$x_1 = \sum_{p=0}^{n}\mu^{\frac{p}{2}}\frac{dS_p}{dy_1}, \quad x_2 = \sum_{p=0}^{n}\mu^{\frac{p}{2}}\frac{dS_p}{dy_2}. \tag{15}$$

3 Second Approximation

As we have seen, these series are divergent, but if we stop at the nth term as we have done in Eq. (15), the residual error can be very small if μ is very small, as I presented above.

We have seen that the quantity called α above is always zero. Another proof of this essential fact can be given.

Set:

$$T = S_1 + \mu S_3 + \mu^2 S_5 + \cdots \mu^{\frac{p-3}{2}} S_{p-2}, \xi = \eta_0 + \mu \eta_2 + \mu^2 \eta_4 + \cdots.$$

I first state that T is a periodic function of y_1 and y_2.

In fact its derivatives dT/dy_1 and dT/dy_2 are periodic functions; therefore we have:

$$T = \beta y_1 + \gamma y_2 + T',$$

where β and γ are constants and T' is a periodic function of y_1 and y_2.

From this we conclude that:

$$\frac{dT}{dy_1} = \beta + \frac{dT'}{dy_1}, \quad \frac{dT}{dy_2} = \gamma + \frac{dT'}{dy_2},$$

where dT'/dy_1 and dT'/dy_2 are trigonometric series where the constant term is zero.

But since the functions $S_1, S_3, \ldots S_{p-2}$ have an odd index, their derivatives change sign on changing from y_2 to $y_2 + 2\pi$. Therefore dT/dy_1 and dT/dy_2 change sign when y_2 increases by 2π. Therefore the constant terms β and γ are zero. Therefore $T = T'$ is a periodic function which does not change when y_1 increases by 2π and which changes sign when y_2 increases by 2π.

Having set that out, we know that ζ_1 and ζ_2 are related by the equation:

$$F(\zeta_1, \zeta_2, y_1, y_2) = C.$$

From this it follows that, if the two values of ζ_2 are merged, then the two values of ζ_1 are also merged.

Writing that the two values of ζ_2 are merged, it follows that:

$$\frac{dT}{dy_2} = 0. \tag{16}$$

This Eq. (16) is furthermore identical to Eq. (14). Now write that the two values of ζ_1 are merged, it will follow that:

$$\frac{dT}{dy_1} + \mu^{\frac{p-1}{2}} \xi = 0. \tag{17}$$

Equations (16) and (17) will have to be equivalent. Additionally they will have to be equivalent to the following:

$$y_2 = \theta(y_1, \mu),$$

were θ has the same meaning as above. Assume that θ is expanded in increasing powers of μ, it will follow:

$$y_2 = \mu\theta_1 + \mu^2\theta_2 + \mu^3\theta_3 + \cdots, \tag{18}$$

where $\theta_1, \theta_2, \theta_3, \ldots$ are periodic functions of y_1.

Assume that y_2 is related to y_1 by Eq. (18); when y_1 increases by 2π, y_2 will not change and T, which is periodic, will not change either; therefore we will have:

$$\int_{y_1=0}^{y_1=2\pi} dT = \int_0^{2\pi} \left(\frac{dT}{dy_1} dy_1 + \frac{dT}{dy_2} dy_2 \right) = 0,$$

or by replacing dT/dy_1 and dT/dy_2 by their values drawn from Eqs. (16) and (17)

$$-\mu^{\frac{p-1}{2}} \int_0^{2\pi} \xi \, dy_1 = 0.$$

If in

$$\xi = \eta_0 + \mu\eta_2 + \mu^2\eta_4 + \cdots$$

y_2 is replaced by its value (18), it follows:

$$\xi = \xi_0 + \mu\xi_1 + \mu^2\xi_2 + \cdots,$$

where ξ_0, ξ_1, ξ_2, etc., are periodic functions of y_1.

Whatever μ, it should hold that:

$$\int_0^{2\pi} (\xi_0 + \mu\xi_1 + \mu^2\xi_2 + \cdots) dy_1 = 0$$

and consequently:

$$\int_0^{2\pi} \xi_0 \, dy_1 = 2\pi[\xi_0] = 0$$

3 Second Approximation

It is clear that in order to get ξ_0, it is sufficient to make $y_2 = 0$ in η_0, hence it holds that

$$[\eta_0] = \frac{\alpha}{n_1} \cos \frac{y_2}{2}.$$

It therefore follows

$$\frac{2\pi\alpha}{n_1} = 0$$

or

$$\alpha = 0.$$

Which was to be proved.

4 Third Approximation

We now propose to build exactly our asymptotic surfaces or more accurately their intersection with the surface $y_1 = 0$ which is a contactless surface as we saw above.

In our mode of geometric representation, the periodic solution that we are considering is represented by some closed trajectory curve. This closed curve crosses the surface $y_1 = 0$ at a point that I will represent in the drawing by O'.

Two asymptotic surfaces pass through this closed curve; these two surfaces cross the surface $y_1 = 0$ along two curves that I have shown in the figure with solid lines at $AO'B'$ and $A'O'B$.

I will represent the curve $y_1 = y_2 = 0$ with a dashed line.

Let us resume with the notation from Sect. 1; we will consider the series s_1 and s_2 which enter into Eq. (4) from that section; as in Sect. 1, let s_1^p and s_2^p be the sum of the first p terms of the series s_1 and s_2. We have seen that the equations:

$$x_1 = s_1^p(y_1, y_2), \quad x_2 = s_2^p(y_1, y_2)$$

represent surfaces which differ only slightly from the asymptotic surfaces. These surfaces will cross the surface $y_1 = 0$ along curves whose equations are:

$$y_1 = 0, \quad x_1 = s_1^p(0, y_2), \quad x_2 = s_2^p(0, y_2)$$

and which are shown in the figure with dash-double-dot lines (Figure 3).

Fig. 3 Intersection of asymptotic surfaces with $y_1 = 0$

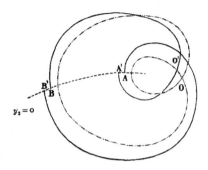

In the preceding section we learned how to form the series s_1 and s_2; we saw that $s_1^p(y_1, y_2)$ and $s_2^p(y_1, y_2)$ are periodic functions with period 2π with respect to y_1 and period 4π with respect to y_2.

It follows from this that the dash-double-dot line must be a closed curve with one double point at O as shown in the figure.

The following is the first question to deal with: are the solid line curves, which are the intersection of the asymptotic surfaces with $y_1 = 0$, also closed curves? It is clear that that is how things would be if the series s_1 and s_2 were convergent. Because the dash-double-dot curves would then differ as little as one wanted from the solid curves, the distance from a point on the solid curve to the dash-double-dot curve would approach 0 when p increased indefinitely.

I am going to prove for a simple example that it is not this way. Let:

$$-F = p + q^2 - 2\mu \sin^2 \frac{y}{2} - \mu\varepsilon \cos x \cdot \varphi(y),$$

where $\varphi(y)$ represents a periodic function of y with period 2π, and where μ and $\in$ are two constants that I assume are very small. I form the equations:

$$\frac{dx}{dt} = -\frac{dF}{dp} = 1, \quad \frac{dy}{dt} = -\frac{dF}{dq} = 2q, \qquad (1)$$

$$\frac{dp}{dt} = \frac{dF}{dx} = -\mu \sin x \cdot \varphi(y), \quad \frac{dq}{dt} = \frac{dF}{dy} = -\mu \sin y + \mu\varepsilon \cos x \cdot \varphi'(y).$$

It can be seen that p and q will play the same role that until now I had given to x_1 and x_2, while x and y will play the role that I had given to y_1 and y_2 the only reason that I changed the notation was to eliminate the indices.

First we assume $\varepsilon = 0$. The equations then allow a periodic solution which is written:

$$x = t, \quad p = 0, \quad q = 0, \quad y = 0.$$

The characteristic exponents (dropping the two which are zero, as always happens with the equations of dynamics) are equal to $\pm\sqrt{2\mu}$.

4 Third Approximation

There are then two asymptotic surfaces whose equations are:

$$p = \frac{dS_0}{dx}, \quad q = \frac{dS_0}{dy}, \quad S_0 = \mp 2\sqrt{2\mu}\cos\frac{y}{2}$$

hence

$$p = 0, \quad q = \pm\sqrt{2\mu}\sin\frac{y}{2}.$$

Since the characteristic exponents are not zero, but instead equal to $\pm\sqrt{2\mu}$ when ε is set equal to 0, there will again exist one periodic solution for small values of ε; two asymptotic surfaces will correspond to this periodic solution. The equation for the asymptotic surfaces can be put in the form

$$p = \frac{dS}{dx}, \quad q = \frac{dS}{dy},$$

where S is a function of x and y which satisfies the equation

$$\frac{dS}{dx} + \left(\frac{dS}{dy}\right)^2 = 2\mu\sin^2\frac{y}{2} + \mu\varepsilon\cos x \cdot \varphi(y).$$

Since the characteristic exponents do not become zero for $\varepsilon = 0$, it follows from what we stated at the end of Sect. 5 in Chap. 3 that p and q and consequently S is expandable in increasing powers of ε. Let us therefore set:

$$S = S_0 + \varepsilon S_1 + \varepsilon^2 S_2 + \cdots$$

We found above that:

$$S_0 = -2\sqrt{2\mu}\cos\frac{y}{2}.$$

As for S_1, it has to satisfy the equation:

$$\frac{dS_1}{dx} + \sqrt{2\mu}\sin\frac{y}{2}\frac{dS_1}{dy} = \mu(\cos x)\varphi(y).$$

If we use Σ to designate a function which satisfies the equation:

$$\frac{d\Sigma}{dx} + \sqrt{2\mu}\sin\frac{y}{2}\frac{d\Sigma}{dy} = \mu e^{ix}\varphi(y)$$

$$\left(i = \sqrt{-1}\right)$$

then S_1 will be the real part of Σ. Now observe that this equation can be satisfied by setting

$$\Sigma = e^{ix}\psi(y);$$

for that it is sufficient that:

$$i\psi + \sqrt{2\mu}\,\sin\frac{y}{2}\frac{d\psi}{dy} = \mu\varphi(y).$$

The equation for ψ that results and which needs to be integrated is linear. Its general integral is written: if $\varphi(y) = 0$

$$\psi = C\left(\tan\frac{y}{4}\right)^{\alpha}, \quad \alpha = -i\sqrt{\frac{2}{\mu}}$$

and if $\varphi(y)$ is arbitrary:

$$\psi = \left(\tan\frac{y}{4}\right)^{\alpha} \int \sqrt{\frac{\mu}{2}}\varphi(y)\left(\sin\frac{y}{2}\right)^{-1}\left(\tan\frac{y}{4}\right)^{-\alpha} dy.$$

Since φ must be expandable in integer powers of y for small values of y, it is easy to see what value will have to be given to the constant of integration. If, for $y = 0$, $\varphi(y)$ is zero, the integral will have to be zero as well, such that it will have to be taken between the bounds 0 and y.

What is necessary now for the curves $BO'B'$ and $AO'A'$ to be closed? The function S has to remain finite and all of its derivatives do too for all values of y and be periodic with period 4π with respect to y (recall that this is what happened for the functions s_1^p and s_2^p which we discussed a little bit ago). Since that should take place for all values of ε, it should occur for S_1, and since S_1 is equal to $\cos x$ multiplied by the real part of ψ, plus $\sin x$ multiplied by the imaginary part of ψ, that should also hold for ψ.

Therefore for the values of y close to 2π, it should be possible to expand ψ according to integer powers of $y - 2\pi$. But it isn't this way for $\left(\tan\frac{y}{4}\right)^{\alpha}$. Therefore the integral:

$$J = \int_0^{2\pi} \sqrt{\frac{\mu}{2}}\varphi(y)\left(\sin\frac{y}{2}\right)^{-1}\left(\tan\frac{y}{4}\right)^{-\alpha} dy$$

should be zero. We calculate this integral by assuming $\varphi(y) = \sin y$. Let us set tangent $\tan\frac{y}{4} = t$, it follows that:

4 Third Approximation

$$J = 4\sqrt{2\mu} \int_0^\infty \frac{t^{-\alpha}(1-t^2)dt}{(1+t^2)^2}.$$

We integrate by parts by observing that $(1-t^2)/(1+t^2)^2$ is the derivative of $t/(1+t^2)$, so it follows:

$$J = 4\alpha\sqrt{2\mu} \int_0^\infty \frac{t^{-\alpha}dt}{1+t^2}.$$

We set $t^2 = u$ and we then have:

$$J = 2\alpha\sqrt{2\mu} \int_0^\infty \frac{u^{-\frac{\alpha+1}{2}}du}{1+u} = \frac{2\pi\alpha\sqrt{2\mu}}{\cos\frac{\alpha\pi}{2}} = \frac{-8\pi i}{e^{\frac{\pi}{\sqrt{2\mu}}} + e^{\frac{-\pi}{\sqrt{2\mu}}}}.$$

Therefore J is not zero; therefore the curves $BO'B'$ and $AO'A'$ are not closed; therefore the series s_1 and s_2 are not convergent, nor are the series defined in Sect. 6 of Chap. 3 and Sect. 3 of this chapter just as I had stated in those sections.

The distance between the two points B and B' is therefore not zero, but it enjoys the following property. Not only does BB' approach 0, when μ approaches 0, but the ratio $BB'/\mu^{p/2}$ also approaches 0 however large p might be.

In fact the equation for the dotted curve is

$$y_1 = 0, \quad x_1 = s_1^p(0, y_2), \quad x_2 = s_2^p(0, y_2)$$

and the equation for the solid curve is

$$y_1 = 0, \quad x_1 = f_1(0, y_2), \quad x_2 = f_2(0, y_2).$$

According to what we saw above the series s_1 and s_2 asymptotically represent the functions f_1 and f_2 which means that:

$$\lim_{\mu \to 0} \frac{f_1 - s_1^p}{\mu^{\frac{p}{2}}} = \lim_{\mu \to 0} \frac{f_2 - s_2^p}{\mu^{\frac{p}{2}}} = 0$$

Therefore the ratio to $\mu^{p/2}$ of the distance between B and the dotted curve will approach 0 and it will also be true of the ratio to $\mu^{p/2}$ of the distance between B' and the dotted curve. It therefore holds that:

$$\lim_{\mu \to 0} \frac{BB'}{\mu^{\frac{p}{2}}} = 0$$

Which was to be proved.

In other words, if we regard μ as an infinitesimal of the first-order, the distance BB', is an infinitesimal of infinite order. It is in this sense that the function $e^{-1/\mu}$ is an infinitesimal of infinite order without being zero.

In the specific example that we dealt with above, the distance BB' is of the same order of magnitude as the integral J, meaning the same order as $e^{-\pi/\sqrt{2\mu}}$.

A second question to be dealt with is to know whether the extension of the two curves $O'B$ and $O'B'$ cross. If that is in fact so, the trajectory which passes by the point of intersection will belong at the same time to two sheets of the asymptotic surface. This will be a *doubly asymptotic* trajectory. Let C be the closed trajectory which passes through the point O' and which represents the periodic solution. The doubly asymptotic trajectory differs very little from C, when t is negative and very large, the trajectory is asymptotically removed from it, separated from it by a lot at first, and then again getting asymptotically closer, so as to differ from C very little when t is positive and very large.

I propose to establish that there are infinitely many doubly asymptotic trajectories.

I start by observing that the curve $O'B$, however far it is extended, will never cross itself, meaning that this extended curve $O'B$ does not have a double point. In fact according to the definition of this curve, the antecedents of various points of $O'B$ are themselves on this curve $O'B$, such that the antecedent of the curve $O'B$ is a portion of this curve. Likewise the second, third and nth antecedent of $O'B$ are smaller and smaller portions of this curve, bounded by the point O' on the one hand and a point D getting closer and closer to O' on the other hand.

If the curve $O'B$ had a double point, all of its antecedents would have to have a double point as well, and consequently so would any arc $O'D$, however small it might be, that is a part of $O'B$. Now, the principles from Sect. 5 of Chap. 3 make it possible for us to construct the portion of $O'B$ near O' and to observe that this curved portion does not have a double point. It therefore follows that it is the same for the entire curve however far it is extended.

According to the definition of the two sheets of the asymptotic surface and of the curves $BO'A'$ and $B'O'A$, one of these curves (for example the curve $BO'A'$) is such that the nth antecedent of a point on this curve gets indefinitely close to O', as n increases; for the other curve $BO'A'$, it is the nth consequent that gets indefinitely close to O'. What we just stated therefore also applies to the curve $O'B'$, provided that the word antecedent is replaced by the word consequent throughout. Therefore the curve $O'B'$ however far it is extended will not cross itself and it is clear that it will also be the same for the curves $O'A$ and $O'A$.

I now state that the curvature of the curves $O'B$ and $O'B'$ is finite; I mean that it does not increase indefinitely when μ approaches 0.

In fact we have seen that not only did the series s_1 and s_2 asymptotically represent the functions f_1 and f_2, but that the series d^2s_1/dy_2^2 and d^2s_2/dy_2^2 asymptotically represent d^2f_1/dy_2^2 and d^2f_2/dy_2^2.

From this it is concluded that if μ is regarded as infinitesimal, the curvature of the solid-line curve at point B will be infinitesimally different from the curvature of the dotted curve at the closest point; now this latter curvature is finite; therefore so is the curvature of the solid-line curve.

Now let B_1 be the consequent of the point B and B_1' that of the point B'. The distance BB_1 is of the same order of magnitude as $\sqrt{\mu}$ and the same is true of the distance $B'B_1'$; the arcs BB_1 and $B'B_1'$ are therefore very small if μ is very small and their curvature is finite. Additionally, the distances BB_1 and $B'B_1'$ and likewise the ratios BB'/BB_1 and $BB'/B'B_1'$ approach 0 when μ approaches 0. Finally, there exists a positive integral invariant.

We now find ourselves in the conditions for Theorem III from Sect. 4 of Chap. 2. From this, we will conclude that the arcs BB_1 and $B'B_1'$ cross, meaning that the curve $O'B'$ crosses the extended curve $O'B$ and consequently that there is at least one doubly asymptotic trajectory.

I now state that there are at least two.

In fact the figure was constructed such that the points B and B' are on the curve

$$y_1 = y_2 = 0$$

But the origin for the y_2 has remained arbitrary; I may assume that it is chosen such that it holds that $y_2 = 0$ at the point of intersection of the two curves $O'B$ and,. In this case, the points B and B' coincide. The same must therefore happen for their consequents B_1 and B_1'. The two arcs BB_1 and $B'B_1'$ therefore have the same ends, but that is not sufficient to satisfy Theorem III that I just applied (in fact in order to satisfy this theorem, the area delimited by these two arcs must not be convex); it is also necessary that they cross at another point N.

Through this point a doubly asymptotic trajectory will pass which is not coincident with the one which passes through B. There are therefore at least two doubly asymptotic trajectories.

I continue to assume that the points B and B' coincide. Let *BMN* be the portion of the curve $O'B$ included between the points B and N; similarly let *BPN* be the portion of the curve $O'B'$ included between the point $B = B'$ and the point N. These two arcs *BMN* and *BPN* will delimit some area that I call α.

We have seen that in the specific case of the three-body problem with which we are occupied one can apply Theorem I from Sect. 4 of Chap. 2. There will therefore exist trajectories which will pass through the area α infinitely many times.

Therefore among the consequents of the area α, there will be infinitely many which will have a portion in common with α.

If therefore the closed curve *BMNPB* which delimits the area α and the consequents of this curve are considered, these consequents which will cross the curve *BMNPB* itself are infinitely many.

How can that happen?

The arc *BMN* cannot cross any of its consequents; because the arc *BMN* and its consequents belong to the curve $O'B$ and the curve $O'B$ cannot re-cross itself.

For the same reason the arc *BPN* cannot cross any of its consequents.

It therefore must be that either the arc *BMN* intersects one of the consequents of *BPN* or else the arc *BPN* crosses one of the consequents of *BMN* (under the assumptions in which we find ourselves, it is the second case that occurs). In one case like the other, the curve $O'B$ or its extension will cross the curve $O'B'$ or its extension.

These two curves therefore cross at infinitely many points and infinitely many of these intersection points are located on the arcs BMN or BPN. Infinitely many doubly asymptotic trajectories will pass through these intersection points.

In the same way one could demonstrate that the asymptotic surface which crosses the surface $y_1 = 0$ along the curve $O'A$ contains infinitely many doubly asymptotic trajectories.

Chapter 6
Various Results

1 Periodic Solutions of the Second Kind

In the previous chapter and in particular in Sects. 2 and 3, we built our series by assuming that C_1 was given a value sometimes greater than and sometimes equal to $-\varphi_4$.

Now let us assume that we had given C_1 a value $< -\varphi_4$. Then

$$x_2^1 = \sqrt{\frac{2}{N}([F_1] + C_1)}$$

is not always real. Assume for example that, for the chosen value of C_1, x_2^1 remains real when y_2 varies from η_5 to η_6. I am going to consider a value η_7 for y_2 included between η_5 and η_6:

$$\eta_5 < \eta_7 < \eta_6$$

and I am going to try to define the x_i^k for all values of y_2 included between η_5 and η_7.

I first observe that x_2^1 could have two equal values of opposite sign because of the two signs of the radical; let us first give for example this radical the $+$ sign.

Imagine that we have successively calculated

$$x_1^1, x_1^2, \ldots x_1^{k-2},$$

$$x_2^1, x_2^2, \ldots x_2^{k-2}.$$

Equation (7) from Sect. 3 of Chap. 5 gives us

$$x_1^2[x_2^{k-1}] = \theta(y_2) + C_{k-1},$$

where $\theta(y_2)$ is a completely known function of y_2 and C_{k-1} is a constant.
We will determine this constant from the following condition:
$$\theta(\eta_5) + C_{k-1} = 0.$$

Now, while x_2^1 becomes zero for $y_2 = \eta_5$, the function
$$[x_2^{k-1}] = \frac{\theta(y_2) - \theta(\eta_5)}{x_2^1}$$

remains finite for $y_2 = \eta_5$.

We have therefore fully determined the functions x_i^k for $\eta_5 < y_2 < \eta_7$ and we will call $x_{0,i}^k$ the functions of y_2 thus determined.

Assume that we start the calculation over by giving the radical the −sign. We will then find new values for the functions x_i^k that I will call $x_{1,i}^k$ and which will furthermore be the analytic continuation of the first ones.

Next imagine that C_1 is replaced by a new constant C_1' very close to C_1.
Then the radical
$$\sqrt{\frac{2}{N}([F_1] + C_1')}$$

will be real every time that y_2 is included between η_7 and some value η_8 very close to η_6.

Having laid that out, using the method presented above we are going to calculate the functions x_i^k for the values of y_2 included between η_7 and η_8, first by making
$$x_2^1 = +\sqrt{\frac{2}{N}([F_1] + C_1')}$$

(we will call the functions calculated in that way $x_{2,i}^k$), and then by making
$$x_2^1 = -\sqrt{\frac{2}{N}([F_1] + C_1')}$$

(we will call the functions calculated in that way $x_{3,i}^k$).

Next we are going to construct the four branches of curves
$$1: y_1 = 0, \ x_1 = \varphi_{0,1}(y_2), \ x_2 = \varphi_{0,2}(y_2)$$

that we will extend from $y_2 = \eta_5$ to $y_2 = \eta_7$.
$$2: y_1 = 0, \ x_1 = \varphi_{1,1}(y_2), \ x_2 = \varphi_{1,2}(y_2)$$

that we will also extend from $y_2 = \eta_5$ to $y_2 = \eta_7$.

1 Periodic Solutions of the Second Kind

$$3 : y_1 = 0, \; x_1 = \varphi_{2,1}(y_2), \; x_2 = \varphi_{2,2}(y_2)$$

that we will also extend from $y_2 = \eta_7$ to $y_2 = \eta_8$.

$$4 : y_1 = 0, \; x_1 = \varphi_{3,1}(y_2), \; x_2 = \varphi_{3,2}(y_2)$$

that we will also extend from $y_2 = \eta_7$ to $y_2 = \eta_8$.

In these formulas we have set

$$\varphi_{p,q}(y_2) = x_{p,q}^o + x_{p,q}^1 \sqrt{\mu} + \cdots x_{p,q}^k \mu^{\frac{k}{2}}.$$

The first and second curves will connect and will be tangent at a single point on the curve $y_2 = \eta_5$.

The third and fourth curves will connect and will be tangent at a single point on the curve $y_2 = \eta_8$.

This is what is shown in Fig. 1 where the three dotted lines represent the three curves

$$y_2 = \eta_5, \eta_7, \eta_8,$$

and where the arc AB represents the first of our four branches of the curve, the arc AD' prime the second, the arc $B'C$ the third, and the arc DC the fourth.

We will regard C_1 as a given, but C_1' until now has remained arbitrary. We will fix C_1' by the condition that the first and third curves connect and that the points B and B' are coincident, which is expressed analytically by the following conditions:

$$\varphi_{0,1}(\eta_7) = \varphi_{2,1}(\eta_7), \quad \varphi_{0,2}(\eta_7) = \varphi_{2,2}(\eta_7). \tag{1}$$

Additionally, these two equations are not distinct and reduce to a single equation.

By relying on theorem III from Sect. 4 of Chap. 2, we will be able to prove that if C_1' is determined by Eq. (1), the equations

$$\varphi_{1,1}(\eta_7) = \varphi_{3,1}(\eta_7), \quad \varphi_{1,2}(\eta_7) = \varphi_{3,2}(\eta_7) \tag{1'}$$

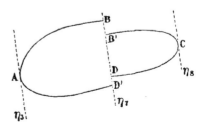

Fig. 1 The three dotted lines represent the three curves

will also be satisfied up to quantities of order $\mu^{(k+1)/2}$; meaning that the second and fourth curves will connect up to quantities of this order, or that the distance DD' is an infinitesimal of the same order as $\mu^{(k+1)/2}$.

But I need to make the same observation here that I made above: the series that one reaches in this way are not convergent although they can provide useful service if handled with precaution.

There are therefore regions, where, at least during some time, y_2 (in the case where it is assumed that $n_2 = 0$) or $n_2 y_1 - n_1 y_2$ (in the general case) retain finite values. This is the fact that astronomers ordinarily express when they state that there is libration. One can ask whether these regions of libration are crisscrossed with periodic solutions.

This can be taken into account by the following considerations.

Let us write the following equations:

$$x_1 = x_1^0 + \mu x_1^2, \quad x_2 = x_2^0 + \sqrt{\mu}\sqrt{\frac{2}{N}([F_1] + C_1)} + \mu u_2^2. \qquad (2)$$

Up to quantities of order μ, these equations are those of surfaces that we have constructed (see Fig. 1); they therefore approximately satisfy Eq. (3) from Sect. 2 of Chap. 5. As for u_2^2, it is a function of y_1 and y_2 which only differs from x_2^2 by a function of y_2 such that

$$\frac{du_2^2}{dy_1} = \frac{dx_2^2}{dy_1}.$$

This function u_2^2 must additionally always remain finite.

I propose to change the form of the function F which enters into our differential equations such that Eq. (2) *exactly* satisfies Eq. (3) from Sect. 2 of Chap. 5.

I am therefore looking for function F^* such that the equations

$$\frac{dx_1}{dt} = \frac{dF^*}{dy_1}, \quad \frac{dx_2}{dt} = \frac{dF^*}{dy_2},$$
$$\frac{dy_1}{dt} = -\frac{dF^*}{dx_1}, \quad \frac{dy_2}{dt} = -\frac{dF^*}{dx_2} \qquad (3)$$

allow trajectory surfaces precisely represented by Eq. (2).

This is how we will determine this function F^*.

Let us first observe that x_1^0 and x_2^0 are determined by the two simultaneous equations:

$$F_0(x_1^0, x_2^0) = C, \quad \frac{dF_0}{dx_2^0} = 0.$$

From these two equations, x_1^0 and x_2^0 can be drawn as functions of C. We will therefore now regard x_1^0 and x_2^0 as known functions of C.

1 Periodic Solutions of the Second Kind

Additionally, $[F_1]$ is a function of y, x_1^0 and x_2^0; this will allow us to view it as a known function of y_2 and C.

Equation (2) will consequently give us x_1 and x_2 as functions of y_1, y_2, C and C_1. Observe that if x_1 and x_2 are defined by these equations, then

$$x_1 dy_1 + x_2 dy_2 = dS$$

is an exact differential such that

$$\frac{dx_1^2}{dy_2} = \frac{du_2^2}{dy_1}.$$

Now let us solve Eq. (2) for C and C_1; it will follow that

$$C = F^*(x_1, x_2, y_1, y_2),$$
$$C_1 = \Phi^*(x_1, x_2, y_1, y_2).$$

In this way, the function F^* is defined and we will have, using Jacobian notation,

$$[F^*, \Phi^*] = 0,$$

which means that

$$\Phi^* = \text{const.},$$

is an integral of Eq. (3).

The most general solution of Eq. (3) is then written as

$$\frac{dS}{dy_1} = x_1, \quad \frac{dS}{dy_2} = x_2, \quad \frac{dS}{dC} = C' + t, \quad \frac{dS}{dC_1} = C_1', \tag{α}$$

where C' and C_1' are two new constants of integration.

We will try to actually form F^* or at least understand the order of magnitude of the difference:

$$F - F^*.$$

Now, x_1^2 is defined by the following condition:

$$F(x_1^0 + \mu x_1^2, x_2^0 + \sqrt{\mu} x_2^1) = C$$

must be a quantity of the same order as $\mu\sqrt{\mu}$ (see Sect. 2 of Chap. 5).

Therefore, since dF_0/dx_2^0 is zero, the function

$$F\left(x_1^0 + \mu x_1^2, x_2^0 + \sqrt{\mu} x_2^1 + \mu u_2^2\right) = C$$

will again be of the same order as $\mu\sqrt{\mu}$, however small the function u_2^2.
Furthermore, it holds identically that

$$F^*\left(x_1^0 + \mu x_1^2, x_2^0 + \sqrt{\mu} x_2^1 + \mu u_2^2\right) = C.$$

Therefore, the difference

$$F - F^*$$

regarded as a function of μ, C, C_1, y_1 and y_2 is of order $\mu\sqrt{\mu}$.

Now we set

$$\xi_2 \sqrt{\mu} = x_2 - x_2^0.$$

From the two Eq. (1), C_1 and C can easily be found as functions of x_1, ξ_2, y_1, y_2 and μ; it can then be seen without difficulty that C and C_1 are expandable in positive powers of $\sqrt{\mu}$, where the coefficients are finite functions of x_1, ξ_2, y_1, and y_2.

We just saw that $F - F^*$ is a function of μ, y_1, y_2, C and C_1 for which the expansion in powers of μ start with the term in $\mu\sqrt{\mu}$; if in the expansion we replace C and C_1 by their values as functions of μ, x_1, ξ_2, y_1, and y_2, we will see that this difference $F - F^*$ is a function expanded in powers of μ whose coefficients depend on x_1, ξ_2, y_1, and y_2 and which starts by a term in $\mu\sqrt{\mu}$.

Consequently, the function

$$\frac{F - F^*}{\mu\sqrt{\mu}} = F'(\mu, x_1, \xi_2, x_1, x_2)$$

does not become infinite for $\mu = 0$.

By change of variable we are now led to make Eq. (3) become

$$\frac{dx_1}{dt} = \frac{dF^*}{dy_1}, \quad \frac{dy_1}{dt} = -\frac{dF^*}{dx_1},$$
$$\frac{d\xi_2}{dt} = \frac{dF^*}{\sqrt{\mu}dy_2}, \quad \frac{dy_2}{dt} = -\frac{dF^*}{\sqrt{\mu}d\xi_2}. \qquad (3')$$

Similarly, the proposed equations

$$\frac{dx_i}{dt} = \frac{dF}{dy_i}, \quad \frac{dy_i}{dt} = -\frac{dF}{dx_i}$$

1 Periodic Solutions of the Second Kind

must reduce to

$$\frac{dx_1}{dt} = \frac{dF}{dy_1}, \frac{dy_1}{dt} = -\frac{dF}{dx_1},$$
$$\frac{d\xi_2}{dt} = \frac{dF}{\sqrt{\mu}dy_2}, \frac{dy_2}{dt} = -\frac{dF}{\sqrt{\mu}d\xi_2}. \tag{4}$$

We will furthermore form the following equations:

$$\frac{dx_1}{dt} = \frac{d}{dy_1}(F^* + \varepsilon F'), \frac{dy_1}{dt} = -\frac{d}{dx_1}(F^* + \varepsilon F'),$$
$$\frac{d\xi_2}{dt} = \frac{d}{\sqrt{\mu}dy_2}(F^* + \varepsilon F'), \frac{dy_2}{dt} = -\frac{d}{\sqrt{\mu}d\xi_2}(F^* + \varepsilon F'), \tag{5}$$

which reduces to (3') for $\varepsilon = 0$ and to (4) for $\varepsilon = \mu\sqrt{\mu}$.

Based on what we saw above, Eq. (3) and consequently Eq. (3') can be integrated exactly; with Eq. (α) we have supplied their general solution.

If we discuss this general solution and if we look at how to build it by keeping the same mode of geometric representation as in the previous sections, we will see that there are infinitely many closed trajectory surfaces.

These surfaces, which have

$$\frac{x_2}{x_1} = \frac{x_2^0 + \sqrt{\mu}\sqrt{\frac{2}{N}([F_1] + C_1)} + \mu u_2^2}{x_1^0 + \mu x_1^2} \tag{6}$$

for their equation, differ slightly from the surfaces that we constructed in Sect. 2 of Chap. 5 and for which the equation was written as

$$\frac{x_2}{x_1} = \frac{x_2^0 + \sqrt{\mu}\sqrt{\frac{2}{N}([F_1] + C_1)}}{x_1^0}. \tag{7}$$

They have the same general shape as the surfaces defined by Eq. (7). If we therefore make the same assumptions as in Sect. 2 of Chap. 5 concerning the maxima and minima of $[F_1]$, two of our surfaces (6) will be closed surfaces with a double curve; these will be the ones which correspond to the values $-\varphi_2$ and $-\varphi_4$ of the constant C_1. The others are made up of one or two closed sheets.

The closed surface with a double curve will be an asymptotic surface for Eq. (3') and the surface will divide the space into three regions as we saw above.

Among these regions, I distinguish the region R_2 included between the two sheets, which is a region of libration, and I propose to show that in this region: infinitely many closed trajectories corresponding to periodic solutions can be traced.

Now in fact return to Eq. (α) that will let us know the general solution of Eqs. (3) and (3′). According to the form of Eq. (2), we can write

$$S = ay_1 + by_2 + \theta(y_1, y_2) + \sqrt{\frac{2\mu}{N}} \int \sqrt{([F_1] + C_1)} dy_2,$$

where a and b are functions of C and C_1 only, and $\theta(y_1, y_2)$ is a real, periodic function of y_1 and y_2.

From this one can deduce that

$$C_1' = \frac{dS}{dC_1} = \frac{da}{dC_1} y_1 + \frac{db}{dC_1} y_2 + \frac{d\theta}{dC_1} + \sqrt{\frac{\mu}{2N}} \int \frac{dy_2}{\sqrt{([F_1] + C_1)}}.$$

We will give C_1 a fixed value which will have to be smaller than $-\varphi_4$ because we are assuming we are located in the region R_2.

Since the closed surface which corresponds to this value of C_1 has the same connections as a torus, we will be able to go around in two different ways: first by regarding y_2 as constant and second by regarding y_1 as constant.

When going around the surface keeping y_2 constant, y_1 will increase by 2π and dS/dC_1 will increase by

$$2\pi \frac{da}{dC_1}.$$

When going around the surface keeping y_1 constant, y_2 will return to the same value, but the integral

$$\int \frac{dy_2}{\sqrt{([F_1] + C_1)}}$$

will have increased by some period v defined as follows:

Assume that the values of y_2 for which the radical $\sqrt{([F_1] + C_1)}$ is real are values included between η_5 and η_6, then it will follow that

$$v = 2 \int_{\eta_5}^{\eta_6} \frac{dy_2}{\sqrt{([F_1] + C_1)}}.$$

When our integral increases by v, dS/dC_1 will increase by

$$v \sqrt{\frac{\mu}{2N}}.$$

1 Periodic Solutions of the Second Kind

For the solution corresponding to this value of C_1 to be periodic, it is necessary and sufficient that these two quantities

$$2\pi \frac{da}{dC_1} \quad \text{and} \quad v\sqrt{\frac{\mu}{2N}}$$

be mutually commensurable.

This condition will obviously be satisfied for infinitely many values of C_1; our region R_2 therefore contains infinitely many closed trajectories representing periodic solutions.

Thus if K is an arbitrary rational number, the equation

$$2\pi \frac{da}{dC_1} = Kv\sqrt{\frac{\mu}{2N}} \tag{8}$$

(which contains C_1 because da/dC_1 and v are functions of C_1) will give us a value of C_1 corresponding to a periodic solution.

In order to discuss this equation, I need to look for what da/dC_1 is.

It is sufficient for me to recall that

$$a = x_1^0 + \mu[x_1^2]$$

and that

$$F_0\left(x_1^0 + \mu x_1^2, x_2^0 + \sqrt{\mu}x_2^1\right) + \mu F_1\left(x_1^0, x_2^0, y_1, y_2\right)$$

must, up to quantities of order $\mu\sqrt{\mu}$, reduce to C. From this it is concluded that

$$-n_1 x_1^2 - \frac{N}{2}\left(x_2^1\right)^2 + F_1\left(x_1^0, x_2^0, y_1, y_2\right) = 0$$

hence

$$-n_1\left[x_1^2\right] - C_1 - [F_1] + [F_1] = 0$$

and

$$\frac{da}{dC_1} = -\frac{\mu}{n_1}.$$

Therefore, da/dC_1 is a constant, independent of C_1, such that Eq. (8) can be written as

$$\frac{v}{\sqrt{\mu}} = \text{constant.} \tag{8'}$$

In order to discuss this new equation, it is appropriate to look for how v varies when C_1 is varied from $-\varphi_4$ to $-\varphi_1$.

For $C_1 = -\varphi_4$, v is infinite; as C_1 varies from $-\varphi_4$ to $-\varphi_1$, v first decreases to some minimum and then subsequently increases again to infinity.

For $C_1 < -\varphi_2$, v can have two values corresponding to the two sheets of the surface and they can be considered separately (see Fig. 7 in Chap. 5).

The first sheet of the surface remains real when C_1 is included between $-\varphi_2$ and $-\varphi_3$; the corresponding value of v decreases from infinity to some minimum when C_1 decreases from $-\varphi_2$ to $-\varphi_3$.

The second sheet of the surface remains real when C_1 is included between $-\varphi_2$ and $-\varphi_1$; the corresponding value of v decreases from infinity to some minimum when C_1 decreases from $-\varphi_2$ to $-\varphi_1$.

Thus, v has at least three minima and always remains greater than some positive bound.

If we regard Eq. (8') as defining C_1 as a function of μ, C_1 will be a continuous function of μ but we could take μ small enough so that this equation does not have any root.

In that way it is certain that infinitely many periodic solutions always exist; but when μ is decreased, all these solutions disappear one after another.

It follows from the preceding that for $\varepsilon = 0$ Eq. (5) allows infinitely many periodic solutions; the principles from Chap. 3 (first part) allow us to affirm that there are still infinitely many for sufficiently small values of ε. Since μ is very small, it seems very likely that there will exist infinitely many periodic solutions for $\varepsilon = \mu\sqrt{\mu}$; meaning for Eq. (4) which are deduced by a very simple change of variables from the proposed equations.

Consequently, if we return to these proposed equations, we see that in the *region of libration R_2 there are infinitely many closed trajectories representing periodic solutions*. We are additionally going to establish this rigorously by a different route.

But if one of these closed trajectories is followed while decreasing μ continuously, then it will also be seen to deform continuously and next *disappear* for some value of μ. In that way for $\mu = 0$, all periodic solutions from the region R_2 will have disappeared one after the other. This is not how the periodic solutions studied in Chap. 3 (first part) behaved; those solution still remained for $\mu = 0$.

It can be shown that in the neighborhood of a closed trajectory representing a periodic solution, either stable or unstable, infinitely many other closed trajectories also pass. This is not sufficient, rigorously, to conclude that any region of the space, however small it might be, is crossed by infinitely many closed trajectories,[1] but it is sufficient to give a high degree of likelihood to this assumption.

---

[1] From Cantor's recent work we have learned in fact (using his language) that a set can be perfect without being continuous.

1 Periodic Solutions of the Second Kind

And so as I just stated, the preceding summary is not sufficient for rigorously establishing the existence of periodic solutions of the second type. Here is how we will arrive at the proof.

Let us again take up our differential equations

$$\frac{dx_i}{dt} = \frac{dF}{dy_i}, \quad \frac{dy_i}{dt} = -\frac{dF}{dx_i}$$

and we will imagine a periodic solution of the first type with period T; when t increases by T, y_1 and y_2 will increase by $n_1 T$ and $n_2 T$; as above I will assume that

$$n_1 T = 2\pi, \quad n_2 = 0.$$

That stated, from the equation

$$F = C$$

we can draw x_2, as a function of x_1, y_1, and y_2; by replacing x_2 by the value thus found, we find

$$\frac{dx_1}{dt} = \frac{dF}{dy_1}, \quad \frac{dy_1}{dt} = -\frac{dF}{dx_1}, \quad \frac{dy_2}{dt} = -\frac{dF}{dx_2},$$

where the right-hand sides can be regarded as known functions of x_1, y_1, and y_2. Finally by eliminating dt it will follow

$$\frac{dx_1}{dy_1} = X, \quad \frac{dy_2}{dy_1} = Y, \tag{9}$$

where X and Y are known functions of x_1, y_1, and y_2 and are periodic with period 2π with respect to y_1.

Let

$$x_1 = \varphi_1(y_1), \quad y_2 = \varphi_2(y_1)$$

be the periodic solution considered which will have period 2π with respect to y_1; I assume that this periodic solution is the one that we defined above and which was shown approximately in the figure from Sect. 2 of Chap. 5 by the isolated closed curve for the surface $C_1 = -\varphi_3$. This will therefore be a *stable* periodic solution; it will have two characteristic exponents, α and $-\alpha$ which are equal and have opposite sign and whose square will be real and negative.

Now let

$$x_1 = \varphi_1(y_1) + \xi_1, \quad y_2 = \varphi_2(y_1) + \xi_2$$

be a slightly different solution from the first. In keeping with the notation from Chap. 3 (first part), let β_1 and β_2 be the initial values of ξ_1 and ξ_2 for $y_1 = 0$, and $\beta_1 + \psi_1$ and $\beta_2 + \psi_2$ be the values of ξ_1 and ξ_2 for $y = 2k\pi$ (where k is an integer). The solution will be periodic with period $2k\pi$ if it holds that

$$\psi_1 = \psi_2 = 0. \tag{10}$$

It is known that ψ_1 and ψ_2 could be expanded in powers of β_1 and β_2 and that these functions will furthermore depend on μ.

If for a moment we look at β_1, β_2, and μ as coordinates of a point in space, then Eq. (10) represents a nonplanar curve and to each point of this nonplanar curve there corresponds a periodic solution. It is clear that ψ_1 and ψ_2 become zero with β_1 and β_2; in fact, if we set $\xi_1 = 0$ and $\xi_2 = 0$, then according to the definition of ξ_1 and ξ_2, we get a periodic solution with period 2π which can also be regarded as periodic with period $2k\pi$.

The curve (10) therefore first includes the entire μ-axis. I propose to prove that if, for $\mu = \mu_0$, $k\alpha$ is a multiple of $2\pi i$, there will exist another branch of the curve (10) which will pass through the point

$$\mu = \mu_0, \ \beta_1 = 0, \ \beta_2 = 0$$

and consequently that for the values of μ near μ_0, there exist other periodic solutions besides $\xi_1 = \xi_2 = 0$.

Set $\mu - \mu_0 = \lambda$ and look to expand ψ_1 and ψ_2 in powers of β_1, β_2 and λ.

We first calculate the terms of first degree in β_1 and β_2.

I first state that for $\lambda = 0$, meaning for $\mu = \mu_0$, all these terms are zero.

In fact, let us assume that the ξ are sufficiently small that their squares can be neglected. We have seen in the first part that in this case, the ξ satisfy a system of two linear differential equations that we have called the perturbation equations of Eq. (9). We have also seen that these linear equations allow two remarkable solutions; when y_1 increases by 2π, the first of these solutions is multiplied by $e^{2\alpha\pi}$ and the other is multiplied by $e^{-2\alpha\pi}$.

For $\lambda = 0$, $k\alpha$ is a multiple of $2\pi i$ such that $e^{2\alpha\pi}$ and $e^{-2\alpha\pi}$ are both kth roots of unity. Therefore, our solutions repeat when y_1 increases by $2k\pi$. Since the equation is linear, the general solution is a linear superposition of these two remarkable solutions, and it too does not change when y_1 increases by $2k\pi$.

Since ψ_1 and ψ_2 are the increases which ξ_1 and ξ_2 undergo when y_1 goes from the value 0 to the value $2k\pi$, ψ_1 and ψ_2 must be zero; but that is only true when ξ_1 and ξ_2 (or β_1 and β_2) are sufficiently small that their squares can be neglected; therefore, only the terms of ψ_1 and ψ_2 which are of first degree in β_1 and β_2 will be zero.

Which was to be proved.

Let

$a\beta_1 + b\beta_2$ be the terms of first degree in ψ_1,

1 Periodic Solutions of the Second Kind

$c\beta_1 + e\beta_2$ be the terms of first degree in ψ_2.

We just saw that for $\lambda = 0$,

$$a = b = c = e = 0.$$

Again for $\lambda = 0$, let

$$\frac{da}{d\lambda} = a', \frac{db}{d\lambda} = b', \frac{dc}{d\lambda} = c', \frac{de}{d\lambda} = e'.$$

I state that

$$a' + e' = 0.$$

In fact, the equation for S

$$(a - S)(e - S) - bc = 0$$

has

$$S = 1 - e^{2ak\pi}, \quad S = 1 - e^{-2ak\pi}$$

as roots (see Sect. 4 of Chap. 3). For $\lambda = 0$, these two roots are zero; if λ is small enough so that the square can be neglected, they will be equal to

$$\pm 2k\pi \frac{da}{d\lambda} \lambda.$$

The equation for S:

$$(a' - S)(e' - S) - b'c' = 0$$

will therefore have two roots

$$S = \pm 2k\pi \frac{da}{d\lambda}$$

and since these two roots are equal and of opposite sign, it must be

$$a' + e' = 0.$$

Additionally a', e', b' and c' will not in general be zero at the same time. In fact that could only happen if $da/d\lambda = da/d\mu$ were zero. Now μ_0 is a quantity selected such that α (which is a function of μ) is commensurable with $2\pi i$. Also $da/d\mu$ could not become zero for all commensurable values of $a/2\pi i$ except by becoming identically zero; hence α would be a constant (which would additionally have to be zero because $\alpha = 0$ for $\mu = 0$) which does not happen in general.

We have seen that for $\lambda = 0$, the terms of first degree in ψ_1 and ψ_2 become zero identically. Let us now assume that the terms of second degree, third degree, up to $m-1$ degree do the same, but that the terms of mth degree in ψ_1 and ψ_2 do not become identically zero for $\lambda = 0$. Let θ_1 be all of the terms in ψ_1 for $\lambda = 0$. Let θ_2 be all of the terms in ψ_2 for $\lambda = 0$. In that way θ_1 and θ_2 are two homogeneous polynomials of mth degree in β_1 and β_2 and of these two polynomials at least one does not become identically zero.

Let us set

$$\psi_1 = a'\lambda\beta_1 + b'\lambda\beta_2 + \theta_1 + \omega_1,$$
$$\psi_2 = c'\lambda\beta_1 - a'\lambda\beta_2 + \theta_2 + \omega_2.$$

Then ω_1 and ω_2 will be a set of terms which will be either at least $m+1$ degree in the β, or else of at least second degree in the β and at least first degree in λ, or else of at least first degree in the β and at least second degree in λ.

I propose to prove that one can draw β_1 and β_2 from Eq. (10) as a series ordered in powers of $\lambda^{\frac{1}{m-1}}$ and for which the terms are not all zero.

But first I need to prove that it holds identically that

$$\frac{d\theta_1}{d\beta_1} + \frac{d\theta_2}{d\beta_2} = 0.$$

In fact there is a positive integral invariant. From this we conclude that there is an integral

$$\iint \Phi(\xi_1, \xi_2) d\xi_1 d\xi_2$$

which has the same value for an arbitrary area belonging to the portion of the contactless surface $y_1 = 0$ and for all its consequents.

Additionally, the function Φ is positive. Therefore $\Phi(0,0)$ is not zero, by multiplying the function Φ by suitable factor, we can always assume that

$$\Phi(0,0) = 1.$$

But the point

$$\xi_1 = \beta_1 + \psi_1, \; \xi_2 = \beta_2 + \psi_2, \; y_1 = 2\pi k$$

is the kth consequent of the point:

$$\xi_1 = \beta_1, \; \xi_2 = \beta_2, \; y_1 = 0.$$

1 Periodic Solutions of the Second Kind

For an arbitrary area it will then hold that

$$\iint \Phi(\beta_1+\psi_1,\beta_2+\psi_2)(d\beta_1+d\psi_1)(d\beta_2+d\psi_2) = \iint \Phi(\beta_1,\beta_2)d\beta_1 d\beta_2$$

from which follows the identity

$$\Phi(\beta_1+\psi_1,\beta_2+\psi_2)\left[\frac{d(\beta_1+\psi_1)}{d\beta_1}\frac{d(\beta_2+\psi_2)}{d\beta_2} - \frac{d(\beta_1+\psi_1)}{d\beta_2}\frac{d(\beta_2+\psi_2)}{d\beta_1}\right]$$
$$= \Phi(\beta_1,\beta_2). \tag{11}$$

We will assume $\lambda = 0$; we will therefore have

$$\psi_1 = \theta_1 + \omega_1, \quad \psi_2 = \theta_2 + \omega_2.$$

The θ then contain only terms of mth degree and the ω only terms of $m+1$ degree or higher.

It results from this that the difference

$$\Phi(\beta_1+\psi_1,\beta_2+\psi_2) - \Phi(\beta_1,\beta_2)$$

only contains terms of at least mth degree in β_1 and β_2. If we agree to neglect terms of mth degree and higher, then it is possible to write

$$\Phi(\beta_1+\psi_1,\beta_2+\psi_2) = \Phi(\beta_1,\beta_2).$$

By continuing to neglect terms of mth degree, we will have

$$\frac{d(\beta_1+\psi_1)}{d\beta_1} = 1 + \frac{d\theta_1}{d\beta_1}, \quad \frac{d(\beta_2+\psi_2)}{d\beta_2} = 1 + \frac{d\theta_2}{d\beta_2},$$
$$\frac{d(\beta_1+\psi_1)}{d\beta_2} = \frac{d\theta_1}{d\beta_2}, \quad \frac{d(\beta_2+\psi_2)}{d\beta_1} = \frac{d\theta_2}{d\beta_1}$$

and

$$\frac{d(\beta_1+\psi_1)}{d\beta_1}\frac{d(\beta_2+\psi_2)}{d\beta_2} - \frac{d(\beta_1+\psi_1)}{d\beta_2}\frac{d(\beta_2+\psi_2)}{d\beta_1} = 1 + \frac{d\theta_1}{d\beta_1} + \frac{d\theta_2}{d\beta_2},$$

such that by identifying all the terms of degree less than m in the identity (11) we arrive at the following relationship:

$$\Phi(\beta_1,\beta_2)\left(\frac{d\theta_1}{d\beta_1} + \frac{d\theta_2}{d\beta_2}\right) = 0.$$

On the left-hand side of this relationship, we will only have to retain the terms of degree at most $m-1$ such that it remains

$$\frac{d\theta_1}{d\beta_1} + \frac{d\theta_2}{d\beta_2} = 0.$$

Which was to be proved.

Let us set

$$\lambda = \pm \eta^{m-1}, \; \beta_1 = \gamma_1 \eta, \; \beta_2 = \gamma_2 \eta;$$

it can be seen that θ_1 and θ_2 become divisible by η^m, and ω_1 and ω_2 by η^{m+1}, such that we can set

$$\theta_1 = \eta^m \theta_1', \; \theta_2 = \eta^m \theta_2', \; \omega_1 = \eta^{m+1} \omega_1', \; \omega_2 = \eta^{m+1} \omega_2',$$

from which it follows

$$\psi_1 = \pm \eta^m (a'\gamma_1 + b'\gamma_2) + \eta^m \theta_1' + \eta^{m+1} \omega_1',$$
$$\psi_2 = \pm \eta^m (c'\gamma_1 - a'\gamma_2) + \eta^m \theta_2' + \eta^{m+1} \omega_2',$$

Equation (10) can be replaced by the following:

$$\begin{aligned} \pm(a'\gamma_1 + b'\gamma_2) + \theta_1' + \eta \omega_1' &= 0, \\ \pm(c'\gamma_1 - a'\gamma_2) + \theta_2' + \eta \omega_2' &= 0. \end{aligned} \quad (12)$$

I state that from these equations γ_1 and γ_2 can be determined as series ordered in powers of η and without the series being identically zero.

In light of Theorem IV, Sect. 2 of Chap. 1 to do that it is sufficient to establish the following:

(1) Equation (12) when we set $\eta = 0$ in them, allow at least one family of real solutions:

$$\gamma_1 = \gamma_1^0, \; \gamma_2 = \gamma_2^0.$$

(2) That if one sets

$$\eta = 0, \; \gamma_1 = \gamma_1^0, \; \gamma_2 = \gamma_2^0,$$

then the functional determinant of the left-hand side of Eq. (12) with respect to γ_1 and γ_2 is not zero.

This amounts to stating that, for Eq. (12) simplified by the assumption that $\eta = 0$, the solution

1 Periodic Solutions of the Second Kind

$$\gamma_1 = \gamma_1^0, \; \gamma_2 = \gamma_2^0$$

must be a single solution.

But because of Theorems V and VI from Sect. 2 of Chap. 1 and their corollaries, γ_1 and γ_2 can still be expanded in powers of η, even if the solution should be multiple, provided that the order of multiplicity is odd.

We are therefore led to consider the following equations:

$$\begin{aligned} \pm(a'\gamma_1 + b'\gamma_2) + \theta_1' &= 0, \\ \pm(c'\gamma_1 - a'\gamma_2) + \theta_2' &= 0 \end{aligned} \tag{13}$$

and we need to look to prove that these equations allow at least one real, odd-order solution.

From these equations, we can draw the following:

$$(a'\gamma_1 + b'\gamma_2)\theta_2' - (c'\gamma_1 - a'\gamma_2)\theta_1' = 0, \tag{14}$$

which is homogeneous and from which one could consequently draw the ratio γ_1/γ_2.

It is clear that θ_1' and θ_2' are formed with γ_1 and γ_2 just as θ_1 and θ_2 are with β_1 and β_2; it follows that

$$\frac{d\theta_1'}{d\gamma_1} + \frac{d\theta_2'}{d\gamma_2} = 0.$$

This proves that there exists a polynomial f which is homogeneous and has degree $m+1$ in γ_1 and γ_2 and which is such that

$$\theta_1' = \frac{df}{d\gamma_2}, \; \theta_2' = \frac{df}{d\gamma_1}.$$

Likewise if we set

$$f_1 = \frac{1}{2}\left(b'\gamma_2^2 + 2a'\gamma_1\gamma_2 - c'\gamma_1^2\right)$$

it follows that

$$a'\gamma_1 + b'\gamma_2 = \frac{df_1}{d\gamma_2}, \; c'\gamma_1 - a'\gamma_2 = -\frac{df_1}{d\gamma_1}$$

such that Eq. (14) can be written as

$$\frac{df_1}{d\gamma_2}\frac{df}{d\gamma_1} - \frac{df_1}{d\gamma_1}\frac{df}{d\gamma_2} = 0.$$

Consider the expression

$$H = \frac{f^2}{f_1^{m+1}}.$$

It is homogeneous and zeroth degree in γ_1 and γ_2; it therefore only depends upon the ratio γ_1/γ_2. I state that the denominator f_1 can never become zero.

In fact the equation

$$(a' - S)(e' - S) - b'c' = 0$$

must have two imaginary roots, hence

$$a'e' - b'c' = -a'^2 - b'c' < 0;$$

hence

$$a'^2 + b'c' > 0,$$

which proves that the quadratic form f_1 is defined. The expression H can therefore never become infinite. It will therefore allow at least one maximum. For this maximum, it must be that

$$\frac{df_1}{d\gamma_2}\frac{df}{d\gamma_1} - \frac{df_1}{d\gamma_1}\frac{df}{d\gamma_2} = 0.$$

In that way, Eq. (14) allows at least one real root. It will in general be single. In any case, it will always be of odd order, because a maximum can only correspond to an odd-order root.

From Eq. (14) we have drawn the ratio γ_1/γ_2; we can therefore set

$$\gamma_1 = \delta_1 u, \quad \gamma_2 = \delta_2 u,$$

where δ_1 and δ_2 are known quantities.

It is now left for us to determine u. In order to do that, I will replace γ_1 and γ_2 by $\delta_1 u$ and $\delta_2 u$ in the first of Eq. (13), it follows that

$$\theta'_1 = u^m \theta''_1, \quad \theta'_2 = u^m \theta''_2,$$

where θ''_1 is formed with δ_1 and δ_2 like θ'_1 with γ_1 and γ_2; hence,

$$\pm(a'\delta_1 + b'\delta_2) + \theta''_1 u^{m-1} = 0. \tag{15}$$

This equation must determine u; if m is even it will have one real root; if m is odd (and this is what will happen in general) it will have two real roots, or no real root; it

1 Periodic Solutions of the Second Kind

will have two real roots if θ_1'' and $\pm(a'\delta_1 + b'\delta_2)$ are of opposite sign; but because of the $\pm$ sign it is always possible to arrange for it to be that way.

Equation (15) therefore allows at least one real root. Additionally, this real root is a single root. There could only be an exception if

$$a'\delta_1 + b'\delta_2 = 0 \quad \text{or} \quad \theta_1'' = 0.$$

But in this case Eq. (15) could be replaced with the following:

$$\pm(c'\delta_1 - a'\delta_2) + \theta_2'' u^{m-1} = 0.$$

There would be no further difficulty unless we had simultaneously

$$a'\delta_1 + b'\delta_2 = c'\delta_1 - a'\delta_2 = 0$$

or else

$$\theta_1'' = \theta_2'' = 0.$$

The first circumstance cannot occur because of the inequality

$$a'^2 + b'c' > 0.$$

The second circumstance could on the other hand come up. It can happen that Eq. (14) allows one root such that $\theta_1' = \theta_2' = 0$. But I state that in this case Eq. (14) will again allow at least one root and for that reason this circumstance will not arise.

In fact it holds identically

$$mf = \gamma_2 \theta_1' - \gamma_1 \theta_2'.$$

If therefore

$$\theta_1' = \theta_2' = 0$$

then it will be that $f = 0$ and because f_1 is never zero, then

$$H = 0.$$

It can in fact happen that the expression H allows 0 as a maximum or minimum. But this expression is not identically zero because θ_1' and θ_2' are not both identically zero; additionally it always remains finite; it therefore becomes either positive or negative; if it becomes positive, it will have a positive maximum which is different from 0; if it becomes negative it will have a negative minimum which is different from 0.

Thus Eq. (14) always allows at least one real root of odd order such that θ_1'' and θ_2'' both do not become zero at the same time.

Therefore, Eq. (13) has at least one real solution of odd order.

Therefore, one can find series which are not identically zero, which are expandable in positive fractional powers of $\mu - \mu_0$ and which satisfy Eq. (10) when they are substituted for β_1 and β_2.

Therefore, there exists a set of periodic solutions with period $2\pi k$ which for $\mu = \mu_0$ are identical with the solution

$$x_1 = \varphi_1(y_1),\ x_2 = \varphi_2(y_1).$$

These are the periodic solutions of the second kind.

2 Divergence of Lindstedt's Series

I would like to finish the presentation of the general results of this monograph by calling particular attention to the negative conclusions which follow from it. These conclusions are very interesting, both because they make clearer the strangeness of the results obtained, but also because it is precisely their negative nature that allows their immediate extension to more general cases; in contrast, the positive conclusions cannot be generalized without specific proof.

I propose to first prove that the series proposed by Lindstedt are not convergent; but before that I want to review what Lindstedt's method consists of. I will present it, it is true, with different notations from that which this worthy astronomer had adopted, because in the interest of greater clarity I want to retain the notation which I used above.

Let us put the equations of dynamics in the same form as in the second part of this monograph and let us write

$$\frac{dx_1}{dt} = \frac{dF}{dy_1},\ \frac{dx_2}{dt} = \frac{dF}{dy_2},\ \frac{dy_1}{dt} = -\frac{dF}{dx_1},\ \frac{dy_2}{dt} = -\frac{dF}{dx_2}.$$

where F will be a given function of the 4 variables x_1, x_2, y_1, and y_2 and we will have

$$F = F_0 + \mu F_1,$$

where F_0 will be a function of x_1 and x_2, independent of y_1 and y_2, and μ will be a very small coefficient such that μF_1 will be the *perturbing function*.

This is in fact the form in which the problems of dynamics and in particular the problems of celestial mechanics present themselves.

If μ were zero, x_1 and x_2 would be constants. If μ is not zero but instead very small, and we call ξ_1 and ξ_2 the initial values of x_1 and x_2, the differences $x_1 - \xi_1$ and $x_2 - \xi_2$ will be the same order of magnitude as μ.

2 Divergence of Lindstedt's Series

If we therefore call n_1 and n_2 the values of $-dF_0/dx_1$ and $-dF_0/dx_2$ for $x_1 = \xi_1$ and $x_2 = \xi_2$, then the differences

$$-\frac{dF_0}{dx_1} - n_1 \text{ and } -\frac{dF_0}{dx_2} - n_2$$

will be the same order of magnitude as μ which will allow us to set

$$-\frac{dF_0}{dx_1} - n_1 = \mu\varphi_1(x_1, x_2),$$
$$-\frac{dF_0}{dx_2} - n_2 = \mu\varphi_2(x_1, x_2),$$

where φ_1 and φ_2 are functions of x_1 and x_2 which are not very large.

The equations of motion are then written as

$$\frac{dx_1}{dt} = \mu\frac{dF_1}{dy_1}, \quad \frac{dx_2}{dt} = \mu\frac{dF_1}{dy_2},$$
$$\frac{dy_1}{dt} = n_1 + \mu\left(\varphi_1 - \frac{dF_1}{dx_1}\right), \quad \frac{dy_2}{dt} = n_2 + \mu\left(\varphi_2 - \frac{dF_1}{dx_2}\right).$$

Now assume that x_1, x_2, y_1, and y_2 instead of regarded directly as functions of t are regarded as functions of two variables:

$$w_1 \text{ and } w_2$$

and that we set

$$w_1 = \lambda_1 t + \varpi_1, \quad w_2 = \lambda_2 t + \varpi_2,$$

where ϖ_1 and ϖ_2 will be arbitrary constants of integration and where λ_1 and λ_2 will be constants that the remainder of the calculation will fully determine.

The equations of motion will then become

$$\lambda_1 \frac{dx_1}{dw_1} + \lambda_2 \frac{dx_1}{dw_2} - \mu\frac{dF_1}{dy_1} = 0,$$
$$\lambda_1 \frac{dx_2}{dw_1} + \lambda_2 \frac{dx_2}{dw_2} - \mu\frac{dF_1}{dy_2} = 0,$$
$$\lambda_1 \frac{dy_1}{dw_1} + \lambda_2 \frac{dy_1}{dw_2} - n_1 - \mu\left(\varphi_1 - \frac{dF_1}{dx_1}\right) = 0,$$
$$\lambda_1 \frac{dy_2}{dw_1} + \lambda_2 \frac{dy_2}{dw_2} - n_2 - \mu\left(\varphi_2 - \frac{dF_1}{dx_2}\right) = 0.$$

(1)

Let us now set

$$
\begin{aligned}
x_1 &= x_1^0 + \mu x_1^1 + \mu^2 x_1^2 + \mu^3 x_1^3 + \cdots, \\
x_2 &= x_2^0 + \mu x_2^1 + \mu^2 x_2^2 + \mu^3 x_2^3 + \cdots, \\
y_1 - w_1 &= \mu y_1^1 + \mu^2 y_1^2 + \cdots, \\
y_2 - w_2 &= \mu y_2^1 + \mu^2 y_2^2 + \cdots, \\
\lambda_1 &= \lambda_1^0 + \mu \lambda_1^1 + \mu^2 \lambda_1^2 + \cdots, \\
\lambda_2 &= \lambda_2^0 + \mu \lambda_2^1 + \mu^2 \lambda_2^2 + \cdots.
\end{aligned} \tag{2}
$$

I assume that the coefficients λ_i^k are constants and that the coefficients y_i^k and x_i^k are trigonometric series ordered according to sines and cosines of multiples of w_1 and w_2.

I will furthermore assume—as I always have up to now—that F_1 is a trigonometric series depending on sines and cosines of multiples of y_1 and y_2 and that the coefficients of this series are mapping functions of x_1 and x_2.

Under these conditions, if I substitute in place of λ_1, λ_2, y_1, y_2, x_1, and x_2 in the left-hand sides of Eq. (1) their values (2), I will have four functions expanded according to increasing powers of μ and it is clear that the coefficients for the various powers of μ will be series ordered according to the trigonometric lines of multiples of w_1 and w_2.

I call these four functions as

$$\Phi_1,\ \Phi_2,\ \Phi_1'\ \text{and}\ \Phi_2'.$$

Having established that this is what Lindstedt's theorem consists of.

It is possible, however, large q, to determine the $2q+2$ constants:

$$\lambda_1^0, \lambda_1^1, \cdots \lambda_1^q,$$
$$\lambda_2^0, \lambda_2^1, \cdots \lambda_2^q,$$

and the $4q$ trigonometric series:

$$x_1^0, \ldots x_1^q,$$
$$x_2^0, \ldots x_2^q,$$
$$y_1^0, \ldots y_1^q,$$
$$y_2^0, \ldots y_2^q,$$

so as to simultaneously make the terms independent of μ and the coefficients of the first q powers of μ become zero in

2 Divergence of Lindstedt's Series

$$\Phi_1, \Phi_2, \Phi'_1, \text{ and } \Phi'_2,$$

so as, in other words, to satisfy the equations of motion up to quantities of order μ^{q+1}.

First we find

$$\lambda_1^0 = n_1,\ \lambda_2^0 = n_2,\ x_1^0 = \xi_1 + \omega_1,\ x_2^0 = \xi_2 + \omega_2,$$

where ω_1 and ω_2 are constants of integration that we will assume to be of order μ.

Let us assume that with a previous calculation we have determined

$$\lambda_i^0, \lambda_i^1, \ldots \lambda_i^{q-1},$$
$$x_i^0, x_i^1, \ldots x_i^{q-1},$$
$$y_i^1, \ldots y_i^{q-1},$$

and that it is now time to determine

$$\lambda_1^q, \lambda_2^q, x_1^q, x_2^q, y_1^q, y_2^q.$$

In order to do that, let us write that the coefficient of μ^q is zero in Φ_1, Φ_2 and Φ'_1, Φ'_2.

It follows that

$$\begin{aligned}
n_1 \frac{dx_1^q}{dw_1} + n_2 \frac{dx_1^q}{dw_2} &= X_1, \\
n_1 \frac{dx_2^q}{dw_1} + n_2 \frac{dx_2^q}{dw_2} &= X_2, \\
n_1 \frac{dy_1^q}{dw_1} + n_2 \frac{dy_1^q}{dw_2} + \lambda_1^q &= Y_1, \\
n_1 \frac{dy_2^q}{dw_1} + n_2 \frac{dy_2^q}{dw_2} + \lambda_2^q &= Y_2,
\end{aligned} \tag{3}$$

where X_1, X_2, Y_1 and Y_2 are known functions.

X_1, X_2, Y_1 and Y_2 are trigonometric series in w_1 and w_2.

In order for the integration of Eq. (3) to be possible, it is necessary that

(1) The ratio n_1/n_2 is not commensurable, which it is always possible to assume.
(2) In the trigonometric series X_1 and X_2, the constant terms are zero. It is always that way, but the proof of this important fact is delicate and does not have a place here; I stop at saying that it must be based on the use of integral invariants.

(3) In the trigonometric series Y_1 and Y_2, the constant terms reduce to λ_1^q and λ_2^q; since λ_1^q and λ_2^q are two unknowns, we will determine these unknowns with this condition.

The integration of Eq. (3) is then possible. Their integration will introduce four arbitrary constants. With each new approximation we will in that way have four more constants of integration; we will give them arbitrary values and the only arbitrary constants that we will retain are ω_1, ω_2, ϖ_1, and ϖ_2.

Thus Lindstedt's series are trigonometric series in w_1 and w_2; they are expanded in powers of μ and also in powers of two constants ω_1 and ω_2.

According to Lindstedt's theorem, these series *formally* satisfy the equations of motion. Therefore, if they were uniformly convergent, they would give us the general integral of these equations.

I state that this is not possible.

In fact let us assume that it is so and that our series converge uniformly for all values of time and for sufficiently small values of μ, ω_1 and ω_2.

It is clear that λ_1 and λ_2 are also series ordered in powers of μ, ω_1 and ω_2. For some values of ω_1 and ω_2, the ratio λ_2/λ_1 is commensurable. The specific solutions which correspond to these values of the constants of integration are then periodic solutions.

We have seen above that any periodic solution allows some number of *characteristic exponents*. Let us look at how these exponents can be calculated when one has the general integral of the given equations.

Let

$$x_1 = \psi_1(t, \omega_1, \omega_2, \varpi_1, \varpi_2), \; x_2 = \psi_2(t, \omega_1, \omega_2, \varpi_1, \varpi_2),$$
$$y_1 = \psi_1'(t, \omega_1, \omega_2, \varpi_1, \varpi_2), \; y_2 = \psi_2'(t, \omega_1, \omega_2, \varpi_1, \varpi_2)$$

be this general integral.

Let us assume that by giving to ω_1, ω_2, ϖ_1, and ϖ_2 the determined ω_1^0, ω_2^0, ϖ_1^0, and ϖ_2^0 values that the functions become periodic in t. In order to get the characteristic exponents of the resulting periodic solution, we will form the 16 partial derivatives as

$$\frac{dx_1}{d\omega_1}, \frac{dx_2}{d\omega_1}, \frac{dy_1}{d\omega_1}, \frac{dy_2}{d\omega_1},$$
$$\frac{dx_1}{d\varpi_1}, \frac{dx_2}{d\varpi_1}, \frac{dy_1}{d\varpi_1}, \frac{dy_2}{d\varpi_1},$$
$$\frac{dx_1}{d\omega_2}, \frac{dx_2}{d\omega_2}, \frac{dy_1}{d\omega_2}, \frac{dy_2}{d\omega_2},$$
$$\frac{dx_1}{d\varpi_2}, \frac{dx_2}{d\varpi_2}, \frac{dy_1}{d\varpi_2}, \frac{dy_2}{d\varpi_2}$$

and in it we will next make

2 Divergence of Lindstedt's Series

$$\omega_1 = \omega_1^0, \ \omega_2 = \omega_2^0, \ \varpi_1 = \varpi_1^0, \ \varpi_2 = \varpi_2^0.$$

Then, for example, $dx_1/d\omega_1$ will take the following form:

$$\frac{dx_1}{d\omega_1} = e^{\alpha_1 t}\theta_1(t) + e^{\alpha_2 t}\theta_2(t) + e^{\alpha_3 t}\theta_3(t) + e^{\alpha_4 t}\theta_4(t),$$

where the α are constants and the θ are periodic functions.

The α are then the desired characteristic exponents. Let us apply this rule to the case we are now working on. We have

$$x_1 = \varphi_1(\omega_1, \omega_2, w_1, w_2)$$

where φ_1 is a periodic function of w_1 and w_2.

It then follows that

$$\frac{dx_1}{d\varpi_1} = \frac{d\varphi_1}{dw_1}, \ \frac{dx_1}{d\omega_1} = \frac{d\varphi_1}{d\omega_1} + \left(\frac{d\lambda_1}{d\omega_1}\frac{d\varphi_1}{dw_1} + \frac{d\lambda_2}{d\omega_1}\frac{d\phi_1}{dw_2}\right)t.$$

The three functions

$$\frac{d\varphi_1}{dw_1}, \ \frac{d\varphi_1}{d\omega_1} \text{ and } \frac{d\lambda_1}{d\omega_1}\frac{d\varphi_1}{dw_1} + \frac{d\lambda_2}{d\omega_1}\frac{d\phi_1}{dw_2}$$

are periodic in w_1 and w_2 and consequently in t.

We could find the analogous expressions for $dx_1/d\varpi_2$ and $dx_1/d\omega_2$.

This proves that the characteristic exponents are zero.

Therefore, *if Lindstedt's series were convergent, all the characteristic exponents would be zero*.

In what case is this so?

We saw above the way in which to calculate the characteristic exponents (Sects. 2 and 4 of Chap. 3).

In the last section, we saw that the characteristic exponents for the equations

$$\frac{dx_i}{dt} = \frac{dF_0}{dy_i} + \mu\frac{dF_1}{dy_i}, \ \frac{dy_i}{dt} = -\frac{dF_0}{dx_i} - \mu\frac{dF_1}{dx_i}$$

could be expanded in powers of $\sqrt{\mu}$; we learned to form the equation which gives the coefficient α_1 of $\sqrt{\mu}$.

Let us review how this equation is formed.

In the section referenced we had set

$$C_{ik}^0 = -\frac{d^2 F_0}{dx_i dx_k}, \ B_{ik}^2 = \frac{d^2 F_1}{dy_i dy_k}.$$

In the second derivatives it is assumed that x_1 and x_2 are replaced by x_1^0 and x_2^0, while y_1 and y_2 are replaced by $n_1 t + \varpi_1$ and $n_2 t + \varpi_2$.[2] C_{ik}^0 is therefore a constant and B_{ik}^2 a periodic function of t. I call b_{ik} the constant term of this periodic function.

Let us next set

$$e_{11} = b_{11} C_{11}^0 + b_{12} C_{21}^0, \quad e_{21} = b_{21} C_{11}^0 + b_{22} C_{21}^0,$$
$$e_{12} = b_{11} C_{12}^0 + b_{12} C_{22}^0, \quad e_{22} = b_{21} C_{12}^0 + b_{22} C_{22}^0.$$

The equation which gives us alpha will then be written as

$$\begin{vmatrix} e_{11} - \alpha_1^2 & e_{12} \\ e_{21} & e_{22} - \alpha_1^2 \end{vmatrix} = 0.$$

In order for this equation to have all zero roots, it is necessary that

$$e_{11} + e_{22} = 0$$

and

$$b_{11} C_{11}^0 + 2 b_{12} C_{12}^0 + b_{22} C_{22}^0 = 0. \tag{4}$$

Now, as I have already proved in the referenced section, we have

$$n_1 b_{11} + n_2 b_{12} = n_1 b_{21} + n_2 b_{22} = 0.$$

Therefore, for the identity (4) to hold, it is necessary either that

$$b_{11} = 0 \tag{5}$$

or else that

$$n_2^2 C_{11}^0 - 2 n_1 n_2 C_{12}^0 + n_1^2 C_{22}^0 = 0. \tag{6}$$

Let us deal with the relationship (5) first. If in the perturbing function F_1 we make

$$x_1 = x_1^0, \ x_2 = x_2^0, \ y_1 = n_1 t + \varpi_1, \ y_2 = n_2 t + \varpi_2,$$

then F_1 will become a periodic function of t. Let us assume that this periodic function is expanded in trigonometric series and let ψ be the constant term; ψ will be a periodic function of ϖ_1 and ϖ_2 and it will follow that

---

[2] It is not necessary here to indicate that these values of x_1^0, x_2^0, n_1 and n_2 are those corresponding to the periodic solution being studied; they are not those which we were using above in the discussion of Lindstedt's method. The ratio n_1/n_2 is therefore commensurable.

2 Divergence of Lindstedt's Series

$$b_{ik} = \frac{d^2\psi}{d\varpi_i d\varpi_k}.$$

We will therefore need to have

$$\frac{d^2\psi}{d\varpi_1^2} = 0. \tag{7}$$

We can always assume that the origin of time was chosen such that ϖ_2 is zero and that ψ is a periodic function of only ϖ_1.

Furthermore, the relationship (7) should be (if Lindstedt's series were to converge) satisfied identically. And in fact if one accepts the convergence of the series, there would in fact be infinitely many periodic solutions corresponding to each commensurable value of the ratio n_1/n_2.

If the relationship (7) is an identity and if ψ is a periodic function, this function will have to reduce to a constant.

Let us look at what that means.

Since the perturbing function F_1 is periodic in y_1 and y_2, it can be written as

$$F_1 = \sum A_{m_1 m_2} \cos(m_1 y_1 + m_2 y_2) + \sum B_{m_1 m_2} \sin(m_1 y_1 + m_2 y_2),$$

where m_1 and m_2 are integers while $A_{m_1 m_2}$ and $B_{m_1 m_2}$ are given functions of x_1 and x_2.

It will then hold that

$$\psi = \sum_S A^0_{m_1 m_2} \cos(m_1 \varpi_1 + m_2 \varpi_2) + \sum_S B^0_{m_1 m_2} \sin(m_1 \varpi_1 + m_2 \varpi_2),$$

where the sum represented by the $\sum_S$ sign extends over all terms such that

$$n_1 m_1 + n_2 m_2 = 0,$$

and $A^0_{m_1 m_2}$, and $B^0_{m_1 m_2}$ represent what $A_{m_1 m_2}$ and $B_{m_1 m_2}$ become when x_1 and x_2 are replaced in them by x_1^0 and x_2^0.

Since the periodic terms must disappear from ψ, it will hold that

$$A^0_{m_1 m_2} = B^0_{m_1 m_2} = 0.$$

Thus the coefficients $A_{m_1 m_2}$ and $B_{m_1 m_2}$ of the expansion of the perturbing function must become zero when x_1 and x_2 in them are given values such that

$$n_1 m_1 + n_2 m_2 = 0.$$

Or better, it must be possible to give the ratio n_1/n_2 commensurable values without introducing secular terms into the perturbing function F_1.

It is clear that this is not so in the specific case of the three-body problem that we have examined and that the ratio of the average movements in it cannot be given a commensurable value without introducing secular terms into the perturbing function.

Now turn to the condition (6) which can be written as

$$\left(\frac{dF_0}{dx_2}\right)^2 \frac{d^2 F_0}{dx_1^2} - 2 \frac{dF_0}{dx_1} \frac{dF_0}{dx_2} \frac{d^2 F_0}{dx_1 dx_2} + \left(\frac{dF_0}{dx_1}\right)^2 \frac{d^2 F_0}{dx_2^2} = 0.$$

It expresses that the curve

$$F_0(x_1, x_2) = \text{const.}$$

has an inflection point at the point $x_1 = x_1^0$, $x_2 = x_2^0$.

Since this condition must be satisfied for all values of x_1^0 and x_2^0 which correspond to a commensurable ratio n_1/n_2, the curve $F_0(x_1, x_2) = \text{constant}$ will have to reduce to a family of straight lines.

It is a specific case that we will set aside, because it is clear that nothing of the kind happens in the three-body problem.

In that way, *in the specific case of the three-body problem that we have studied and consequently also in the general case, Lindstedt's series do not converge uniformly for all values of the arbitrary constants of integration that they contain.*

3 Nonexistence of One-to-One Integrals

Return to our equations of dynamics with two degrees of freedom:

$$\frac{dx_i}{dt} = \frac{dF}{dy_i}, \quad \frac{dy_i}{dt} = -\frac{dF}{dx_i}. \quad (i = 1, 2).$$

These equations allow an integral:

$$F = \text{const.}$$

This integral F is an analytic and one-to-one function of x_1, x_2, y_1, y_2 and μ; it is periodic with period 2π in y_1 and y_2.

3 Nonexistence of One-to-One Integrals

I propose to prove that there does not exist another integral having the same properties.

In fact, let

$$\Phi = \text{const.}$$

be another one-to-one analytic integral in x_1, x_2, y_1, y_2 and μ and periodic in y_1 and y_2.

Let

$$x_1 = \varphi_1(t),\ x_2 = \varphi_2(t),\ y_1 = \varphi_3(t),\ y_2 = \varphi_4(t)$$

be a periodic solution (with period T) of our differential equations.

Let

$$x_1 = \varphi_1(t) + \xi_1,\ x_2 = \varphi_2(t) + \xi_2,\ y_1 = \varphi_3(t) + \xi_3,\ y_2 = \varphi_4(t) + \xi_4.$$

Let β_i be the value of ξ_i for $t = 0$; let $\beta_i + \psi_i$ be the value of ξ_i for $t = T$; we know that the ψ are expandable according to increasing powers of the β. Let us consider the equation in S:

$$\begin{vmatrix} \dfrac{d\psi_1}{d\beta_1} - S & \dfrac{d\psi_1}{d\beta_2} & \dfrac{d\psi_1}{d\beta_3} & \dfrac{d\psi_1}{d\beta_4} \\ \dfrac{d\psi_2}{d\beta_1} & \dfrac{d\psi_2}{d\beta_2} - S & \dfrac{d\psi_2}{d\beta_3} & \dfrac{d\psi_2}{d\beta_4} \\ \dfrac{d\psi_3}{d\beta_1} & \dfrac{d\psi_3}{d\beta_2} & \dfrac{d\psi_3}{d\beta_3} - S & \dfrac{d\psi_3}{d\beta_4} \\ \dfrac{d\psi_4}{d\beta_1} & \dfrac{d\psi_4}{d\beta_2} & \dfrac{d\psi_4}{d\beta_3} & \dfrac{d\psi_4}{d\beta_4} - S \end{vmatrix} = 0.$$

The roots of this equation are equal to

$$e^{\alpha T} - 1,$$

where the α are the characteristic exponents; two of these roots are therefore zero, and in the specific case of the three-body problem that we are dealing with, the other roots must be different from zero.

I first remark that we have

$$\begin{aligned}\dfrac{dF}{dx_1}\dfrac{d\psi_1}{d\beta_i} + \dfrac{dF}{dx_2}\dfrac{d\psi_2}{d\beta_i} + \dfrac{dF}{dy_1}\dfrac{d\psi_3}{d\beta_i} + \dfrac{dF}{dy_2}\dfrac{d\psi_4}{d\beta_i} &= 0, \\ \dfrac{d\Phi}{dx_1}\dfrac{d\psi_1}{d\beta_i} + \dfrac{d\Phi}{dx_2}\dfrac{d\psi_2}{d\beta_i} + \dfrac{d\Phi}{dy_1}\dfrac{d\psi_3}{d\beta_i} + \dfrac{d\Phi}{dy_2}\dfrac{d\psi_4}{d\beta_i} &= 0.\end{aligned} \quad (1)$$

$$(i = 1, 2, 3, 4)$$

In the derivatives of F and Φ, x_1, x_2, y_1 and y_2 must be replaced by $\varphi_1(T), \varphi_2(T), \varphi_3(T)$ and $\varphi_4(T)$.

From which, it can be concluded either that

$$\frac{\frac{dF}{dx_1}}{\frac{d\Phi}{dx_1}} = \frac{\frac{dF}{dx_2}}{\frac{d\Phi}{dx_2}} = \frac{\frac{dF}{dy_1}}{\frac{d\Phi}{dy_1}} = \frac{\frac{dF}{dy_2}}{\frac{d\Phi}{dy_2}} \tag{2}$$

or else that the functional determinant of the ψ with respect to the β is zero and that all the sub-determinants of first-order are also.

Additionally by designating the derivative of $\varphi_i(t)$ by $\varphi_i'(t)$, it holds that

$$\frac{d\psi_i}{d\beta_1}\varphi_1'(0) + \frac{d\psi_i}{d\beta_2}\varphi_2'(0) + \frac{d\psi_i}{d\beta_3}\varphi_3'(0) + \frac{d\psi_i}{d\beta_4}\varphi_4'(0) = 0, \tag{3}$$
$(i = 1, 2, 3, 4)$

$$\frac{dF}{dx_1}\varphi_1'(0) + \frac{dF}{dx_2}\varphi_2'(0) + \frac{dF}{dy_1}\varphi_3'(0) + \frac{dF}{dy_2}\varphi_4'(0) = 0,$$
$$\frac{d\Phi}{dx_1}\varphi_1'(0) + \frac{d\Phi}{dx_2}\varphi_2'(0) + \frac{d\Phi}{dy_1}\varphi_3'(0) + \frac{d\Phi}{dy_2}\varphi_4'(0) = 0. \tag{4}$$

From a very simple calculation whose details will be found later, it can be concluded from these equations that if Eq. (2) is not satisfied,
then either

$$\varphi_1'(0) = \varphi_2'(0) = \varphi_3'(0) = \varphi_4'(0) = 0; \tag{5}$$

or else the equation in S will have three zero roots (the other four roots should also be zero, because the characteristic exponents are pairwise equal and of opposite sign).

Hence, we know that the equation in S has only two zero roots; additionally Eq. (5) can only be satisfied for certain very specific periodic solutions (I say that for those who have studied Book X, Chap. 6 of Laplace's Celestial Mechanics) and where the third body describes a circumference like the first two.

Equation (2) will therefore need to be satisfied. They will have to be for

$$x_1 = \varphi_1(T),\ x_2 = \varphi_2(T),\ x_3 = \varphi_3(T),\ x_4 = \varphi_4(T).$$

But since the origin of time has remained arbitrary, they will also need to be satisfied whatever t, for

$$x_1 = \varphi_1(t),\ x_2 = \varphi_2(t),\ x_3 = \varphi_3(t),\ x_4 = \varphi_4(t).$$

In other words, they will be satisfied for all points of all periodic solutions. I now state that these equations are satisfied identically. For example, let us set

3 Nonexistence of One-to-One Integrals

$$f = \frac{d\Phi}{dx_2}\frac{dF}{dx_1} - \frac{d\Phi}{dx_1}\frac{dF}{dx_2}.$$

It is clear that f will again be an analytic and one-to-one function; it will hold that $f = 0$ for all points of all periodic solutions. I want to establish that f is identically zero; in order to do that I am going to show that for $\mu = 0$, it holds identically that

$$0 = f = \frac{df}{d\mu} = \frac{d^2 f}{d\mu^2} = \cdots.$$

In fact let us consider an arbitrary periodic solution of the first kind; let

$$x_1 = \varphi_1(t, \mu), \; x_2 = \varphi_2(t, \mu), \; y_1 = \varphi_3(t, \mu), \; y_2 = \varphi_4(t, \mu)$$

be this solution; it will be possible to expand the functions φ in powers of μ and when μ approaches 0, they will respectively approach

$$x_1^0, \; x_2^0, \; n_1 t + \varpi_1, \; n_2 t + \varpi_2.$$

(where x_1^0 and x_2^0 are constants such that n_1/n_2 is commensurable and ϖ_1 and ϖ_2 are the quantities defined in Sect. 3 of Chap. 3). As long as μ is not zero, it will hold that

$$f(\varphi_1(t, \mu), \varphi_2(t, \mu), \varphi_3(t, \mu), \varphi_4(t, \mu)) = 0.$$

But since the function f is analytic and consequently continuous, it will again hold for $\mu = 0$ (although for $\mu = 0$ the characteristic exponents become zero):

$$f\left(x_1^0, x_2^0, n_1 t + \varpi_1, n_2 t + \varpi_2\right) = 0.$$

But if one considers an arbitrary set of values of x_1 and x_2, one will always be able to find a system x_1^0 and x_2^0 that will differ from it as little as one wishes and which will correspond to a rational value of n_1/n_2. Then let it be

$$\frac{n_1}{n_2} = \frac{\lambda_1}{\lambda_2},$$

where λ_1 and λ_2 are two relatively prime integers. We will choose t such that

$$n_1 t + \varpi_1 = y_1 + 2k\pi.$$
(k integer)

One will then have

$$n_2 t + \varpi_2 = \frac{\lambda_2}{\lambda_1}(y_1 + 2\pi k - \varpi_1) + \varpi_2.$$

If we set

$$n_2 t + \varpi_2 = y_2^0 + 2k'\pi.$$
(k' integer)

one should get

$$f(x_1^0, x_2^0, y_1^0, y_2^0) = 0.$$

Given an arbitrary value of y_2, the integers k and k' can be chosen such that the difference $y_2 - y_2^0$ is smaller in absolute value than $2\pi/\lambda_1$. But we can always choose x_1^0 and x_2^0 in the way that this set of values differs as little as one wants from x_1 and x_2, and that the ratio n_1/n_2 while still being commensurable is such that the integer λ_1 is as large as one wants. Consequently, given an arbitrary set of values x_1, x_2, y_1, and y_2, one will be able to find a set of values which differs from it as little as one would like and for which f will be zero. Since the function f is analytic, it will therefore have to be identically zero for $\mu = 0$.

Having said that and since

$$f(\varphi_i(t, \mu)) = 0$$

whatever t and μ are; it follows, for all points of the periodic solution,

$$\frac{df}{d\mu} + \frac{df}{dx_1}\frac{d\varphi_1}{d\mu} + \frac{df}{dx_2}\frac{d\varphi_2}{d\mu} + \frac{df}{dy_1}\frac{d\varphi_3}{d\mu} + \frac{df}{dy_2}\frac{d\varphi_4}{d\mu} = 0.$$

This relation will be true in particular for

$$\mu = 0, \; x_1 = x_1^0, \; x_2 = x_2^0, \; y_1 = n_1 t + \varpi_1, \; y_2 = n_2 t + \varpi_2.$$

But, when μ is zero, f is identically zero, and consequently its derivatives with respect to the x_i and the y_i are zero. Hence it follows

$$\frac{df}{d\mu} = 0$$

for $\mu = 0$, $x_i = x_i^0$ and $y_i = n_i t + \varpi_i$; and from this one would conclude as above that $df/d\mu$ is identically zero for $\mu = 0$.

In the same way, one could demonstrate the $d^2 f/d\mu^2$ and the other derivatives of f with respect to μ are zero for $\mu = 0$.

3 Nonexistence of One-to-One Integrals

Therefore, the function f is identically zero and Eq. (2) is an identity.

But, if that is so, that means that Φ is a function of F, and that the two integrals Φ and F are not distinct.

Our equations therefore do not have any other analytic and one-to-one integral besides $F = \text{constant}$.

When I state that these equations do not allow any other one-to-one integral, I do not mean to state just that they do not have an integral which remains analytic and one-to-one for all values of x, y and μ.

I mean to state that aside from the integral F, these equations do not allow an integral which remains analytic, one-to-one (and periodic in y_1 and y_2) for all values of y_1 and y_2 and for sufficiently small values of μ when x_1 and x_2 cover an arbitrary domain, however small this domain might furthermore be.

It is known that Bruns proved that apart from the known integrals, the three-body problem does not allow an algebraic integral. This result is thus seen to be matched by an entirely different route.

I stated above that Eqs. (1), (3), and (4) necessarily led to one of the following three consequences: either Eq. (2) is satisfied, or else Eq. (5) is or else the equation in S has at least three zero roots.

In fact, let us form the following 4 row and 5 column matrix:

$$\left\| \frac{d\psi_i}{d\beta_1} \frac{d\psi_i}{d\beta_2} \frac{d\psi_i}{d\beta_3} \frac{d\psi_i}{d\beta_4} \varphi'_i(0) \right\|. \qquad (6)$$
$$(i = 1, 2, 3, 4)$$

If Eqs. (1) and (4) are satisfied without Eq. (2) being satisfied, we have to conclude that all determinants obtained by eliminating two columns and one row from this matrix are zero.

If, now, a linear transformation is applied to x_1, x_2, y_1 and y_2, the ψ and the β will undergo the same linear transformation and the matrix (6) can be simplified.

One can always assume that this linear transformation was chosen such that

$$\frac{d\psi_i}{d\beta_k} = 0 \text{ for } i < k.$$

Then the three-by-three products of the four quantities

$$\frac{d\psi_1}{d\beta_1}, \frac{d\psi_2}{d\beta_2}, \frac{d\psi_3}{d\beta_3}, \frac{d\psi_4}{d\beta_4}$$

are all zero, and hence it follows that at least two of these quantities are zero. We can always assume that the linear transformation was chosen such that it is $d\psi_3/d\beta_3$ and $d\psi_4/d\beta_4$ which are zero.

If furthermore one of the two quantities $d\psi_1/d\beta_1$ and $d\psi_2/d\beta_2$ is also zero, then the equation for S will have three zero roots.

If on the other hand neither of these two quantities is zero, it can be concluded from Eq. (3) that

$$\varphi_1'(0) = \varphi_2'(0) = 0.$$

By eliminating the 3rd and 4th column and the 3rd row from matrix (6) or else the 3rd and 4th column and the 4th row, it follows

$$\frac{d\psi_1}{d\beta_1} \frac{d\psi_2}{d\beta_2} \varphi_3'(0) = \frac{d\psi_1}{d\beta_1} \frac{d\psi_2}{d\beta_2} \varphi_4'(0) = 0,$$

which can only happen either if

$$\varphi_1'(0) = \varphi_2'(0) = \varphi_3'(0) = \varphi_4'(0) = 0,$$

meaning if Eq. (5) is satisfied; or else if

$$\frac{d\psi_1}{d\beta_1} = 0 \text{ or } \frac{d\psi_2}{d\beta_2} = 0,$$

meaning if the equation in S has three zero roots.
Which was to be proved.

Chapter 7
Attempts at Generalization

1 The N-Body Problem

With the preceding results, there is room to hope that they can be extended to the case where the equations of dynamics have more than two degrees of freedom and consequently to the general case of the n-body problem?

It is possible, but it will not happen without a new effort.

I believed when I started this work that once the solution of the problem was found for the specific case that I dealt with it would be immediately generalizable without having to overcome any new difficulties outside of those which are due to the larger number of variables and the impossibility of a geometric representation. I was mistaken.

Also I think that I need to insist a bit here on the nature of the obstacles facing this generalization.

If there are p degrees of freedom, the state of the system can be represented by the position of a point in a $2p - 1$ dimensional space. Most of the conclusions from the first part are still true and do not require any change. There exist therefore infinitely many periodic solutions represented by closed trajectories and which are classified as stable and unstable or even in many more categories according to the nature of their characteristic exponents. There are also infinitely many asymptotic solutions.

I also sought to extend the calculation from Sect. 2 of Chap. 5 to the general case by setting aside the question of convergence. The series obtained in that way can in fact, even though they diverge, be useful for astronomers in certain cases and perhaps guide mathematicians towards the final solution.

Let us assume 3 degrees of freedom and take up Eq. (1) from Sect. 3 of Chap. 3 by making the same assumptions as in that section.

Let us next look for 3 functions of y_1, y_2 and y_3:

$$x_1 = \Phi_1(y_1, y_2, y_3),$$
$$x_2 = \Phi_2(y_1, y_2, y_3),$$
$$x_3 = \Phi_3(y_1, y_2, y_3),$$

satisfying the equations

$$\frac{dx_1}{dy_1}\frac{dF}{dx_1} + \frac{dx_1}{dy_2}\frac{dF}{dx_2} + \frac{dx_1}{dy_3}\frac{dF}{dx_3} + \frac{dF}{dy_1} = 0,$$
$$\frac{dx_2}{dy_1}\frac{dF}{dx_1} + \frac{dx_2}{dy_2}\frac{dF}{dx_2} + \frac{dx_2}{dy_3}\frac{dF}{dx_3} + \frac{dF}{dy_2} = 0,$$
$$\frac{dx_3}{dy_1}\frac{dF}{dx_1} + \frac{dx_3}{dy_2}\frac{dF}{dx_2} + \frac{dx_3}{dy_3}\frac{dF}{dx_3} + \frac{dF}{dy_3} = 0,$$

or, what amounts the same thing, the equations

$$F = C, \quad \frac{dx_1}{dy_2} = \frac{dx_2}{dy_1}, \quad \frac{dx_3}{dy_2} = \frac{dx_2}{dy_3}, \quad \frac{dx_1}{dy_3} = \frac{dx_3}{dy_1}.$$

Let us now assume that x_1, x_2 and x_3 are expandable in powers of μ or $\sqrt{\mu}$ and that for $\mu = 0$, they reduced to the constants x_1^0, x_2^0 and x_3^0.

We next set as above

$$\frac{dF_0}{dx_1^0} = -n_1, \quad \frac{dF_0}{dx_2^0} = -n_2, \quad \frac{dF_0}{dx_3^0} = -n_3.$$

If there is no linear relation with integer coefficients between n_1, n_2 and n_3, then x_1, x_2 and x_3 are expandable in powers of μ; each term is at the same time periodic in y_1, y_2 and y_3. But it introduces small divisors.

If among n_1, n_2 and n_3 there is one and only one linear relationship with integer coefficients

$$m_1 n_1 + m_2 n_2 + m_3 n_3 = 0,$$

the calculations can continue absolutely as in Sect. 3 of Chap. 5. The three functions x_1, x_2, and x_3 are expanded in powers of $\sqrt{\mu}$ and they are at least doubly periodic; I mean that they do not change when y_1, y_2 and y_3 increase by a multiple of 2π such that $m_1 y_1 + m_2 y_2 + m_3 y_3$ does not change; there are again small divisors.

1 The N-Body Problem

There remains a third case, the most interesting of all, which is the one where there are two linear relationships with integer coefficients between n_1, n_2 and n_3

$$m_1 n_1 + m_2 n_2 + m_3 n_3 = 0,$$
$$m'_1 n_1 + m'_2 n_2 + m'_3 n_3 = 0.$$

It is then possible to expand x_1, x_2 and x_3 in powers of $\sqrt{\mu}$ and such that these functions are periodic; I mean that they do not change when y_1, y_2 and y_3 increases by a multiple of 2π and such that $m_1 y_1 + m_2 y_2 + m_3 y_3$ and $m'_1 y_1 + m'_2 y_2 + m'_3 y_3$ do not change. There are no longer small divisors, but the calculation of these functions is not without some difficulties.

In a first approximation, the determination of these functions depends on the integration of a system of differential equations which have the canonical form of the equations of dynamics, *but with only two degrees of freedom*. In nearly all applications, these equations will depend on a very small parameter in which the expansion can be done so as to be able to apply to them the conclusions from Chaps. 1 and 2 (Part I).

In the following approximations, all that will be left to do is to integrate the areas.

That is not all; the problem of n-bodies presents special difficulties that are not found in the general case. Without doubt these difficulties are not as intrinsic as those whose existence I indicated above and with some close attention it should be possible to triumph over them.

But I have to say a few words about them here.

In the n-body problem, F_0 does not depend on all the linear variables x_i; consequently, not only is the Hessian of F_0 with respect to the variables x_i zero, but the Hessian of an arbitrary function of F_0 is also zero (see page 121). This follows because if $\mu = 0$, meaning in Keplerian movement, then the perihelions are fixed.

This difficulty did not exist in the case we covered (first example, Sect. 1 of Chap. 4) because we did not take g, the longitude of perihelion, for the variable, but instead $g - t$. Also, it would not exist with a law of attraction other than Newtonian.

Here are what the strange consequences of this are

We have seen that there are two kinds of periodic solutions: solutions of the first kind, which we spoke about in Chap. 3 (Part I) and which continue to exist however small μ is; and solutions of the second kind which we spoke about in Sect. 1 of Chap 6 and which disappear one after the other when μ is decreased.

In the case of the three-body problem if we set $\mu = 0$, the orbits of the two small bodies reduce to two Keplerian ellipses. What happens to the periodic solutions of the first kind when one sets $\mu = 0$? In other words, what are the periodic solutions of the equations of Keplerian motion? Some correspond to the case where the two mean motions are commensurable. But it is the others whose sight is more unsettling and on which I have to insist.

If $\mu = 0$, it means that the masses of the two planets are infinitesimal mass and that they cannot act on each other in any meaningful manner, *unless at an infinitesimal distance from each other*. But if these planets pass infinitesimally close to each other, their orbits are going to be violently changed as if they collided. Initial conditions can be arranged such that these collisions occur periodically and in that way discontinuous solutions result which are true periodic solutions of the problem of Keplerian motion and *which we are not allowed to set aside.*

Such are the reasons why I have given up, at least momentarily, extending the results obtained to the general case. Not only do I not have enough time, but I think that such an attempt would be premature.

In fact, I have not been able to do a sufficiently in depth study with even the specific case to which I limited myself. It is only after significant research and effort that mathematicians will fully know this field, where I have only done a simple reconnaissance, and when they will find there solid ground from which they will be able to launch new conquests.

Erratum

Page 166, line 4, *instead of*

$$+ \left[(x_1^1)^2 \frac{d^2 F_0}{(dx_1^0)^2} + 2x_1^1 x_2^1 \frac{d^2 F_0}{dx_1^0 dx_2^0} + (x_2^1)^2 \frac{d^2 F_0}{(dx_2^0)^2} \right]$$

read

$$+ \frac{1}{2} \left[(x_1^1)^2 \frac{d^2 F_0}{(dx_1^0)^2} + 2x_1^1 x_2^1 \frac{d^2 F_0}{dx_1^0 dx_2^0} + (x_2^1)^2 \frac{d^2 F_0}{(dx_2^0)^2} \right].$$

Bibliography

Bohlin, K., *Über die Bedeutung des Principe der lebendigen Kraft für die frage von der stabilität dynamisher system*, Acta Mathematica, vol. 10, p. 109–30, 1887.

Bohlin, K., *Zur Frage der Convergenz der Reihenentwickelungen in der Störungstheorie*, Astronomische Nachrichten, vol. 121, num. 2882, columns 18–24, 1889.

Cauchy, A.-L., *Mémoire sur un théorème fondamental, dans le calcul intégral*, Comptes Rendus des séances de l'académie des sciences, vol. 14, p. 1020, 1842.

Hill, G. W., *Researches in the Lunar Theory*, American Journal of Mathematics, vol. 1, p. 5–26, 1878.

Jacobi, C.G.J., *Vorlesungen über Dynamik*, G. Reimer, Berlin, 1884.

Jordan, C., *Cours d'Analyse de l'École polytechnique: Calcul intégral, équations différentielles. Tome troisième*, Gauthier-Villars et fils, Paris, 1887.

Kowalevski, S. *Zur theorie der partiellen Differentialgleichungen*, Journal für die reine und angewandte Mathematik, vol. 80, p. 1–23.

Laplace, P. S., *Traité de Mécanique Céleste, Vol. 5*, Bachelier, Paris, 1825.

Lindstedt, A., *Beitrag zur Integration der Differentialgleichungen des Störungstheorie*, Mémoires de l'Académie Impériale des Sciences de Saint-Pétersbourg, vol. 31, num. 4, p. 1–21, 1883.

Picard, M. E., *Sur la forme des intégrales des équations différentielles du second ordre dans le voisinage de certains points critiques*, Comptes rendus des séances de l'académie des sciences, vol. 20, p. 743–45, 1878.

Poincaré, H., *Note sur les propriétés des fonctions définies par les équations différentielles*, Journal de l'École polytechnique, vol. 45, p. 13–26, 1878.

Poincaré, H., *Sur les propriétés des fonctions définies par les équations aux différences partielles*, Ph.D. thesis, Université de Paris, Gauthier-Villars, Paris, 1879.

Poincaré, H., *Sur certaines solutions particulières du problème des trois corps*, Bulletin astronomique, vol. 1, p. 65–74, 1884.

Poincaré, H., *Sur les courbes définies par les équations différentielles*, Journal de mathématiques pures et appliquées, vol. 2, p. 151–217, 1886.

Poincaré, H., *Sur un moyen d'augmenter la convergence des séries trigonométriques*, Bulletin astronomique, vol. 3, p. 521–28, 1886.

Poincaré, H., *Sur les intégrales irrégulières des équations linéaires*, Acta Mathematica, vol. 8, p. 295–344, 1886.

Puiseux, V., *Recherches sur les Fonctions Algébriques*, Journal de mathématiques pures et appliquées, vol. 15, p. 365–480, 1850.

Tisserand, M. F., *Sur la commensurabilité des moyens mouvements dans le système solaire*, Comptes rendus des séances de l'académie des sciences, vol. 104, pp. 259–65, January 31, 1887.

Author Index

B
Bohlin, K., 66
Bruns, H., 36, 235

C
Callandreau, P.J., 36
Cantor, G., 212
Cauchy, A.-L., 7, 8, 10, 11, 14, 15, 18, 72

F
Floquet, G., 36

G
Gyldén, H., xx

H
Hill, G.W., 63–66

K
Kowalevski, S., 18, 19

L
Laplace, P. S., vi
Lindstedt, A., xx, 69, 222, 224, 226, 229, 230

P
Phragmén, L., xix
Picard, M.E., 31
Poisson, S., 40, 58, 65, 67
Puiseux, V., 16

S
Stieltjes, T., 36

T
Tisserand, M.F., 151

W
Weierstrass, K., 8

Subject Index

B
Body, 11, 46, 51, 58, 66, 68, 104, 106, 150, 152, 164, 169, 201, 230–232, 235, 237, 239

C
Canonical, 4, 103, 169
Canonical form, 103, 104, 155, 158, 159, 166, 239
Characteristic exponents, 85, 86, 88, 89, 107, 108, 115, 120, 125, 127–129, 133, 137, 196, 197, 213, 226, 227, 231–233, 237
Coefficient, periodic, 30–32, 34, 79, 84, 87, 119, 120, 129, 136, 138
comet, xvi
Commensurable, 58, 90, 94, 98, 117, 157, 164, 165, 167, 176, 211, 215, 225, 226, 229, 230, 233, 234, 239
Convergent, 8, 31, 32, 34, 35, 88, 122–124, 127, 128, 133–135, 140, 196, 199, 206, 222, 226, 227
Curve
  asymptotic, 125
  invariant, 45, 46, 70, 74

D
Degrees of freedom, 4, 6, 89, 95, 99, 108, 116, 135, 147, 161, 171, 230, 237, 239
Determinant, 15–17, 38, 39, 48, 53, 80, 83, 84, 88, 94, 95, 100, 101, 108–110, 112, 116, 131, 155, 218, 232, 235
Divergent, 133, 134, 138, 144, 161, 162, 181, 193

E
Earth, 150
Ecliptic, 150

Energy
  conservation, 4, 6, 40, 41, 147, 159
  potential, 40
  total, 4, 47, 95, 158, 164, 168
Equations of dynamics, 4, 46, 50, 51, 86, 89, 103, 123, 127, 147, 158, 159, 161, 196, 222, 230, 237, 239
Expandable, 7, 9–17, 19, 20, 24, 27, 29, 31, 71–73, 81, 90, 95, 105, 108, 109, 111, 112, 115, 119, 125–129, 133, 135, 138, 139, 141, 164, 165, 176, 180, 181, 183, 197, 198, 208, 222, 231, 238

F
Function
  periodic, 20, 27, 30–32, 34, 77–79, 82, 84–86, 90, 93, 96–102, 107, 110, 114, 118–122, 130, 132, 137, 138, 151, 161, 166, 167, 172, 176, 177, 179–184, 188, 189, 191, 193, 194, 196, 210, 227–229
  perturbing, 151, 152, 222, 228–230

G
Gravity
  center of, 23, 42, 66, 104, 150

H
Hessian (determinant), 94, 95, 100

I
Incompressible, 45
Integral, 4, 6, 19–21, 24–27, 29, 30, 39–46, 48, 51, 53–58, 61–63, 65–70, 74, 75, 82, 85, 86, 88, 91, 95, 102, 103, 105, 147, 151, 156, 157, 168, 188, 192, 198, 200, 201, 207, 210, 216, 225, 226, 230, 231, 235

Integration, constant of, 3, 6, 9, 26, 30, 34, 90, 99, 100, 102, 103, 107, 120, 168, 172, 198, 207, 223, 225, 226, 230
Invariant, 43–47, 51, 53–58, 62, 63, 66, 68–70, 74, 86, 156, 157, 201, 216, 225
   curve, 45, 47, 69, 70, 74
Invariant, integral, 43, 45, 47, 53–58, 62, 63, 66, 68, 74, 86, 156, 201, 216

J

Jacobian (determinant), 15, 17, 53, 80, 88, 94, 155
Jupiter, 150

L

Libration, 206, 209, 212
Lunar. *See* Moon

M

Mean anomaly, 105, 151
Mean motion, 58, 151, 157, 239
Moon, 63–65, 150
Motion, laws of, 44, 45, 56

O

Orbit
   eccentricity, 63, 64, 150–152
   inclination, 63, 64, 150
   Keplerian, 66, 239, 240
Osculator major axis, 151

P

Period, 7, 20, 31, 32, 77, 78, 80, 82–86, 90, 91, 93–98, 103, 104, 110, 117, 119, 124, 132, 147, 151, 161, 166, 179, 182, 188, 189, 191, 196, 198, 210, 213, 222, 230, 231
Periodic. *See* coefficient, function, solution
Perturbation equations, 4, 37, 84, 86, 87, 107, 119, 214
Point, 4, 5, 7, 10–12, 23, 31, 40, 41, 44–46, 56–58, 60, 61, 63, 64, 66, 68, 70–75, 80, 81, 94, 124, 147–150, 154–156, 158, 159, 164, 169, 170, 173, 174, 196, 200, 201, 214, 230, 237
Poisson's theorem, 40
Probability, 60, 61

R

Recurrence, 56, 57, 59, 60, 70, 71, 74
Region, 5, 58, 60, 61, 63, 206, 209, 212
Root
   double, 34, 35

   multiple, 80, 81, 117

S

Series
   Fourier, 114
   sines and cosines, 100, 181
   trigonometric, 34, 35, 87, 100, 101, 122, 167, 181, 193, 224, 225, 228
Series expansion, 58, 84
Singularity, 5, 11, 12, 152
Singular points, 12
Small divisors, 238
Small parameter (μ), 152, 164
Solution
   asymptotic, 58, 124, 127, 133, 135, 161, 174, 237
   doubly-asymptotic, 200
   periodic, 7, 77–82, 84, 85, 87, 88, 91, 94, 95, 107, 111, 117, 119, 124, 127, 133, 135, 163, 164, 174, 182, 196, 200, 206, 209, 211–214, 222, 226, 229, 232–234, 237, 239
Stability
   asymptotic, 58
   Poisson, 58
   secular, 87
   temporary, 87, 88
Stability coefficients, 87, 88
State, 6, 7, 9, 12, 24, 45–47, 54, 56, 59, 60, 69, 71, 85, 87, 89, 115, 124, 138, 140, 142, 147, 149, 150, 152, 155, 158, 159, 161, 164, 169, 200, 206, 214, 218, 235, 237
Stirling series, 134, 135, 144
Sun, 63, 64, 150
Surface
   asymptotic, 125, 161–163, 181, 192, 195, 197, 200, 202, 209
   contactless, 56, 68–71, 73, 170, 195, 216
   trajectory, 57, 157, 164, 165, 168, 170, 206, 209

T

Theory of linear substitutions, 33
Trajectory
   closed, 7, 164, 195, 200, 212
   doubly asymptotic, 200, 201
Transformation, 55, 56, 63, 116, 156, 159, 166, 190, 235

V

Volume, 45, 46, 58–61, 68

CPSIA information can be obtained
at www.ICGtesting.com
Printed in the USA
LVHW051400280419
615854LV00006B/246/P